DELTA FUNCTIONS:
Introduction to Generalised Functions
Second Edition

"The dissenting opinions of one generation become the prevailing interpretation of the next"
Hendrick J. Burton (1871-1949)

ABOUT THE AUTHOR

Roland (Roy) Frank Hoskins (as he is known by his colleagues and friends) is a leading authority on Delta Functions. He graduated from Birkbeck College, London University with a BSc. in mathematics to which he added a M.Sc. in the same subject in 1959 with a dissertation on potential theory. He became interested in mathematical analysis while working as an industrial physicist in the 1950's at the Woolwich laboratories of Siemens Brothers in London. His next industrial move was to the Telecommunications Division of Associated Electrical Industries, where from 1959 until 1968 he worked as a mathematician on problems in this field of network analysis and feedback amplifier design. As a result he developed a strong research interest in linear systems theory and in the application of generalised functions to this and allied areas.

The interest continued when in 1968 he became a lecturer in the Department of Mathematics at Cranfield Institute of Technology (now Cranfield University), eventually becoming Professor of Applicable Mathematics. Roy Hoskins is a Fellow of the Institute of Mathematics and its Applications, and was, until recently, Research Professor in the Department of Mathematical Sciences at De Montfort University, Leicester.

He has published numerous papers on electrical networks, linear systems theory, generalised functions and nonstandard analysis. The latter interest led to writing his *Standard and Nonstandard Analysis* (Ellis Horwood Ltd.), *Theories of Generalised Functions: Distributions, Ultradistributions and other Generalised Functions* with co-author J. Sousa Pinto (Horwood Publishing Ltd.) and the predecessor book *Delta Functions: an Introduction to Generalised Functions* (Horwood Publishing Ltd.) of which this book is an updated and extended second edition incorporating the developments of Nonstandard Analysis and Schwartz distributions.

DELTA FUNCTIONS:
An Introduction to Generalised Functions
Second Edition

R. F. HOSKINS
Research Professor
Department of Mathematical Sciences
De Montfort University
Leicester

WOODHEAD
PUBLISHING

Oxford   Cambridge   Philadelphia   New Delhi

Published by Woodhead Publishing Limited,
80 High Street, Sawston, Cambridge CB22 3 HJ, UK
www.woodheadpublishing.com

Woodhead Publishing, 1518 Walnut Street, Suite 1100, Philadelphia,
PA 19102-3406, USA

Woodhead Publishing India Private Limited, G-2, Vardaan House, 7/28 Ansari Road,
Daryaganj, New Delhi – 110002, India
www.woodheadpublishingindia.com

First published by Horwood Publishing Limited, 1999
Second edition 2009
Reprinted by Woodhead Publishing Limited, 2011

British Library Cataloguing in Publication Data
A catalogue record for this book is available from the British Library

ISBN 978-1-904275-39-8

Cover design by Jim Wilkie

Preface

This book is an updated and restructured version of the text first published in 1979 under the title **Generalised Functions** in the Ellis Horwood series, Mathematics and its Applications. The original aim was to give a simple and straightforward, but reasonably comprehensive, account of the theory and application of the delta function, together with an elementary introduction to other, less familiar, types of generalised functions. It was hoped that the book could be used as a working manual in its own right, as well as serving as a preparation for the study of more advanced treatises. The treatment was confined to functions of a single real variable, and little more than a standard background in calculus was assumed throughout the greater part of the text. The attention was focussed primarily on techniques, with a liberal selection of worked examples and exercises.

A revised and extended second edition, with the new and perhaps more appropriate title of **Delta Functions**, was published in 1999, by Horwood Publishing. The central theme remained the delta function of Dirac, its definition, properties and uses. But as well as a brief introduction to the orthodox general theory of Schwartz distributions the revised text included a simplifed account of the recently developed theory of **Nonstandard Analysis** and of the possible interpretation of delta functions and other generalised functions within this context. No other text offering such an approach was readily available at the time, and there seems to have been none issued since which covers such material at a similarly elementary level.

The present edition follows the same general plan in so far as the elementary introductory treatment of delta functions is concerned, but with such modifications as have proved desirable for greater clarity in addressing the conceptual difficulties often experienced by students meeting delta functions for the first time. As before, such results from elementary real

i

analysis as are needed in the main text are summarised in Chapter 1, and the study of generalised functions proper begins with the introduction of the delta function itself in Chapter 2. Chapter 3 then develops the properties of the delta function and its derivatives on a heuristic basis, but with due attention to the dangers involved in an uncritical and too free use of the conventional symbolism. The object is to provide a fairly comprehensive toolkit of rules for the manipulation of algebraic expressions involving delta functions and derivatives of delta functions. Chapter 4 is concerned with applications to linear systems theory, while Chapters 5 and 6 deal respectively with the associated theories of Laplace and Fourier transforms. In particular the definition of the Laplace transforms of the delta function and its derivatives, and the complications which can arise in this context, have been given some extended discussion.

The last five chapters of the book are different in character. Hitherto the aim has been to make the student familiar with the practical use of those forms of generalised functions which are likely to occur most frequently in elementary applications. These include the delta function itself, derivatives of delta functions and infinite periodic sums of delta functions. Chapter 7 introduces other examples of generalised functions which are less well known to students of engineering and physics, but which do have immediate application to an increasing variety of practical problems. This chapter is motivated by the problem of generalising differential and integral operators to fractional order. An outline of the essential features of the standard theory of Schwartz distributions then follows in Chapter 8. It is hoped that this will prepare the reader for the more rigorous and mathematically demanding aspects of the subject to be found in specialised texts on distributions. The material has again been somewhat extended here so as to allow discussion of the problem of defining a comprehensive theory of the distributional Fourier transform, and to indicate that generalised functions exist which are not dstributions in the sense of Schwartz. Chapter 9 contains a similarly brief and elementary outline of what is essentially the theory of the Lebesgue-Stieltjes integral and of the more immediately useful aspects of measure theory.

Finally, in Chapters 10 and 11, there is an elementary introduction to the concepts and methods of Nonstandard Analysis (NSA), with specific application to nonstandard interpretation of delta functions, and of Schwartz distributions in general. NSA appeared in the 1960s, following the remarkable discovery by Abraham Robinson that mathematical analysis could be developed rigorously and simply in terms of a number system containing infinitesimals and infinite elements. Since the previous edition of the present book was published, a number of excellent introductory texts on NSA have become generally available and there is perhaps less justification for the missionary zeal with which Chapters 10 and 11 have been written. Nevertheless it seems to be still widely believed that the prime value of NSA is to be found in the teaching of elementary calculus from a Leibnitzian point of view, with a free but rigorous use of infinitesimal arguments. In fact NSA is a powerful tool which can be applied in virtually all areas of mathematics. Its application to the theory of generalised functions is peculiarly appropriate and needs to be more widely known. Much of the mystery and conceptual difficulty associated with the delta function and other generalised functions is removed when a nonstandard interpretation is offered.

This book and its predecessor arose from courses given to mixed audiences of engineering and mathematics students at Cranfield University and, later, at De Montfort University. My thanks are again due to those students, past and present, whose reactions and comments helped to shape it into its present form. I would also like to thank Professor Jonathan Blackledge for his encouragement and his advice over the preparation of this text and of its predecessor. Finally my thanks are due to Horwood Publishing on the one hand and to my wife on the other for their patience and forbearance during its preparation.

Contents

Chapter 1

Results from Elementary Analysis

1.1   THE REAL NUMBER SYSTEM

1.1.1 Introduction. The term 'analysis' which appears in the heading of this chapter deserves some preliminary comment. It may loosely be described as that branch of mathematics which is concerned with the concepts and processes which underlie the calculus, more especially with the nature of number, function and limit. A clear understanding of such concepts is necessary for the study of advanced calculus and the theory of functions of a real variable. Inevitably this involves re-examining many of the ideas and arguments which are usually taken as intuitively obvious in a first course in calculus, and many students (particularly those whose primary interest is in mathematics as a tool) find this exacting and difficult to justify. One such student was once heard to give a disparaging definition of analysis as 'a sort of revision course in calculus, only done much more slowly'.

It is not our purpose to offer such a course here. Indeed, little more than a standard background in the techniques of elementary calculus is required for the greater part of this book, and most of the material in this chapter should already be familiar to the reader. However the so-called **generalised functions** which are introduced in the main body of the text do require some extension of those techniques and of the concepts which are involved, and it is therefore desirable to begin with some analysis of the standard background itself. In particular this is the

1

case with respect to the structure and properties of the number system which lies at the base of the calculus: that is, the system $\mathbb{R}$ of the **real numbers**. Hence we give first a brief review of the most important features of the real number system.

1.1.2 Algebraic structure of $\mathbb{R}$. Algebraically the real numbers form a structure called a **field**. That is to say, the two fundamental operations of addition and multiplication on $\mathbb{R}$ satisfy the following familiar conditions:

F1. Addition and multiplication are associative and commutative,

$$a + (b + c) = (a + b) + c \quad ; \quad a \times (b \times c) = (a \times b) \times c,$$

$$a + b = b + a \quad ; \quad a \times b = b \times a.$$

F2. There is a distributive law,

$$a \times (b + c) = (a \times b) + (a \times c).$$

F3. There is an additive unit 0 such that $a + 0 = 0 + a = a$, for every element a of $\mathbb{R}$.

F4. There is a multiplicative unit 1 such that $a \times 1 = 1 \times a = a$ for every non-zero element a of $\mathbb{R}$.

F5. For every element a of $\mathbb{R}$ there exists an additive inverse element $(-a)$ such that $a + (-a) = (-a) + a = 0$.

F6. For every non-zero element a of $\mathbb{R}$ there exists a multiplicative inverse element a^{-1} such that $a \times a^{-1} = a^{-1} \times a = 1$.

Next, there is an ordering relation $\leq$ on $\mathbb{R}$ which is compatible with the field operations in the sense that

OF1. $a \leq b$ always implies $a + c \leq b + c$, for every c in $\mathbb{R}$.

OF2. $0 \leq a$ and $0 \leq b$ always implies $0 \leq a \times b$.

These properties show that $\mathbb{R}$ is an **ordered field**, but this is not enough in itself to characterise the real number system uniquely. For example, the system $\mathbb{Q}$ of all rational numbers is itself an ordered field which is properly contained in $\mathbb{R}$.

1.1.3 The Archimedean Property. The real number system $\mathbb{R}$ (and similarly the rational number system $\mathbb{Q}$) is an ordered field which is **Archimedean** in the sense that it satisfies the following property;

A. Let a and b be any two positive real numbers. Then we can always find a positive integer n such that $a < nb$.

This deceptively simple statement has far-reaching consequences. It is obviously equivalent to the fact that if r is any positive real number whatsoever then we can always find a positive integer n which exceeds r. And this in turn is equivalent to saying that there can exist no real number ε such that

$$0 < \varepsilon < 1/n, \qquad \text{for all } n \in \mathbb{N}.$$

For if such an 'infinitely small' (but non-zero) element ε were to exist in $\mathbb{R}$ then its reciprocal, $1/\varepsilon$, would be a real number which is greater than every positive integer, contradicting (A). Such infinitely small numbers, called **infinitesimals**, were postulated and freely used by mathematicians such as Leibniz in the early development of the calculus. Later on, however, infinitesimals fell into disfavour among most mathematicians, although physicists and engineers have continued to make use of them, albeit informally and unsystematically. It was even said that the idea of an infinitesimal was inherently self-contradictory. This last remark is quite unjustified, and in Chapter 10 of this text the status and value of the infinitesimal will be re-evaluated. All that needs to be said here is that the classical number systems such as that of the real numbers $\mathbb{R}$, or of the rational numbers $\mathbb{Q}$, are Archimedean ordered fields and therefore do not contain any non-zero infinitesimal elements.

1.1.4 **Bounds.** Let A be any non-empty subset of $\mathbb{R}$. If there exists a real number M such that $a \leq M$ for every $a \in A$ then the set A is said to be **bounded above** and M is called an **upper bound** of A.

The fundamental property which characterises the real number system can be stated as follows:

B. **Least Upper Bound Property.** Let A be any non-empty set of real numbers which is bounded above. Then there exists a number M_0 which is such that

(i) M_0 is an upper bound of A, and

(ii) if M is any upper bound of A then $M \geq M_0$.

M_0 is said to be the **least upper bound** (l.u.b.) or the **supremum** (sup) of the set A, and we write

$$M_0 = \text{l.u.b.}(A) = \sup(A).$$

There are similar definitions for **bounded below, lower bound** and
greatest lower bound (g.l.b.) or **infimum** (inf), and a corresponding
Greatest Lower Bound property could be stated for the real number
system.

In the usual geometric interpretation of $\mathbb{R}$ the real numbers are identified with, or considered to be labels for, the points of an ideal infinite
straight line. It is the least upper bound property of $\mathbb{R}$ which corresponds
to the fact that the points of the line form a **continuum**, without gaps.
In the case of the rational number system $\mathbb{Q}$ it is easy to see that the
least upper bound property is not universally true. The set of all rational
numbers r such that $r^2 < 2$, for example, is certainly bounded above in
$\mathbb{Q}$ but no rational number exists which is the least upper bound of that
set. Hence the rational numbers alone are not sufficient to label all the
points of the ideal straight line of geometric intuition.

1.1.5 Infinity symbols. It is often convenient to make use of the conventional symbols $+\infty$ and $-\infty$, but it must be stressed that they do
not form part of the real number field itself. They are ideal elements
which may sometimes be adjoined to $\mathbb{R}$ in order to allow more succinct descriptions of regions which are unbounded, or of variables which
increase without limit. The result is a set $\bar{\mathbb{R}} = \mathbb{R} \cup \{-\infty, +\infty\}$, called
the **extended real number system**, which may be equipped with the
following supplementary conventions for algebraic operations and order:

$$(+\infty \times t) = +\infty, \quad (0 < t \le +\infty)$$

$$(+\infty \times t) = -\infty, \quad (-\infty \le t < 0)$$

$$(-\infty \times t) = -\infty, \quad (0 < t \le +\infty)$$

$$(-\infty \times t) = +\infty, \quad (-\infty \le t < 0)$$

$$t/+\infty = t/-\infty = 0 \; ; \quad (-\infty < t < +\infty)$$

$$+\infty + t = +\infty, \quad (t > -\infty) \; ; \quad -\infty + t = -\infty, \quad (t < +\infty).$$

Combinations like $(\pm\infty) + (\mp\infty)$, $(\pm\infty/\pm\infty)$, and $(\pm\infty/\mp\infty)$ are left
undefined.

1.2   FUNCTIONS

1.2.1 Unless stated otherwise we always take the term 'function' to mean a real, single-valued function of a single real variable (usually understood to be time, t). The range of values of the variable for which the function is defined is called the **domain of definition** of the function. Normally we use single letters such as f, g, h to denote functions, and understand an expression like $f(t)$ to mean the value which the function f assumes at the point, or instant, t. For example, we may write $f \geq g$ to indicate that, for every value of t, the number $f(t)$ is always greater than or equal to the corresponding number $g(t)$:

$f \geq g$ if and only if $f(t) \geq g(t)$ for every value of t at which the two functions are defined.

However, it is often inconvenient to adhere strictly to this convention. In cases where the meaning is fairly obvious from the context the expressions f and $f(t)$ may be used more loosely. For example it is common practice to write t^2 to denote both the number obtained by squaring a given number t and also the function which this operation effectively defines.

1.2.2 Sometimes, though not always, a function is defined by a formula valid for every t within the domain of definition. On the other hand it may happen that the formula becomes meaningless for certain values of t, and it is necessary to complete the definition of the function by assigning specific values to $f(t)$ at such points. Usually we are interested in cases when $f(t)$ is defined for all real numbers t, or else when $f(t)$ is defined for all t in a certain finite interval. In the first place we write

$$y = f(t) \ \text{ for } \ -\infty < t < +\infty.$$

In the second it is necessary to distinguish between **open** and **closed** intervals:

$y = f(t)$ for $a < t < b$, f is defined on the open interval (a, b),

$y = f(t)$ for $a \leq t \leq b$, f is defined on the closed interval $[a, b]$.

Examples.

(1) $f(t) = t/|t|$. This formula assigns a well-defined real number to each given non-zero value of t. When $t = 0$, however, the formula reduces to

the expression $0/0$ which has no meaning. To obtain a function whose domain of definition is the entire real axis we would need to specify the value $f(0)$. The most 'natural' way to complete the definition of f would be to write

$$f(t) = t/|t| \text{ for all } t \neq 0, \quad f(0) = 0.$$

Note, however, that other definitions are perfectly possible; if necessary we could always define another function, say g, by writing

$$g(t) = t/|t| \text{ for all } t \neq 0; \quad g(0) = 1.$$

(2) $f(t) = 1/t^2$. Once again we have a function which is well-defined everywhere except at the origin, but this time f(t) assumes arbitrarily large (positive) values as t approaches 0. In such a case it is often convenient to complete the definition by allowing $+\infty$ as a value and setting $f(0) = +\infty$. Technically this gives a function defined on the real number system $\mathbb{R}$ but taking values in the extended real number system $\bar{\mathbb{R}}$.

(3) $f(t) = +\sqrt{1 - t^2}$. Here the domain of definition is the closed interval $[-1, +1]$. For all values of t outside this interval, the formula fails to specify any (real) value.

1.2.3 If, for some finite number M, we have $f(t) \leq M$ for all t in a certain set A of real numbers, then the set

$$f(a) = \{f(t) : \ t \in A\}$$

is a set of real numbers which is bounded above, and which therefore has a least upper bound, M_0. The function f is then itself said to be bounded above over A, and M_0 is called the least upper bound, or supremum, of f. We write

$$M_0 = \text{l.u.b.}_{t \in A}[f(t)], \quad \text{or} \quad M_0 = \sup_{t \in A}[f(t)].$$

Similarly if, for some finite number m, we have $f(t) \geq m$ for all t in a certain range A then f is said to be bounded below over that range and m is called a lower bound of f. The largest number m_0 such that this inequality holds is called the greatest lower bound, or the infimum of f. We write

$$m_0 = \text{g.l.b.}_{t \in A}[f(t)], \quad \text{or} \quad m_0 = \inf_{t \in A}[f(t)].$$

1.3 CONTINUITY

1.3.1 A function f is said to be **continuous** at a point t_0 within its domain of definition if, for all points sufficiently close to t_0, the functional value $f(t)$ differs from $f(t_0)$ by an arbitrarily small amount. More precisely, let any small, positive number ε be given: then we can always find a corresponding positive number η such that whenever $|t - t_0| < \eta$ we must have $|f(t) - f(t_0)| < \varepsilon$.

Equivalently we could say that f is continuous at t_0 if and only if

(i) $f(t_0)$ exists (i.e. t_0 is within the domain of definition of f),

(ii) $f(t)$ tends to a definite limit as t tends to t_0, and

(iii) $\lim_{t \to t_0} f(t) = f(t_0)$.

A function f is said to be **continuous in the open interval** (a, b) if it is continuous at each point t of (a, b). If, in addition, $f(t)$ approaches the value $f(a)$ as its limit as t approaches a, and approaches the value $f(b)$ as t approaches b, then f is said to be **continuous on the closed interval** $[a, b]$.

1.3.2 Let f be a function which is defined and continuous on a finite, closed interval $[a, b]$. Then the following propositions are true for f:

(i) **Intermediate value theorem.** Suppose that $f(a) < f(b)$ and let c be any number lying between $f(a)$ and $f(b)$. Then there exists some point t_0 in the open interval (a, b) such that

$$f(t_0) = c.$$

(ii) If M and m denote respectively the least upper bound and the greatest lower bound of the function f on $[a, b]$, then there exist points t_1 and t_2 in $[a, b]$ such that

$$f(t_1) = M \quad \text{and} \quad f(t_2) = m.$$

This means that for continuous functions we can replace the terms 'least upper bound' and 'greatest lower bound' by 'greatest value of $f(t)$' and 'least value of $f(t)$' respectively. Sometimes it is convenient to emphasise this point by using the notation

$$\max_{a \le t \le b}[f(t)] \quad \text{instead of} \quad \text{l.u.b.}_{a \le t \le b}[f(t)],$$

and
$$\min_{a \leq t \leq b}[f(t)] \quad \text{instead of} \quad \text{g.l.b.}_{a \leq t \leq b}[f(t)].$$

Note that if f fails to be continuous at any point in $[a, b]$, or if we drop either the requirement that $[a, b]$ be a finite interval, or that it be closed, then we can no longer guarantee the existence of a maximum or a minimum value of $f(t)$ in that interval.

(iii) The function f is not merely continuous at each point of $[a, b]$ but is actually **uniformly continuous** on that interval. This means that given any positive number ε, however small, we can always find a corresponding positive number η such that $|f(t) - f(\tau)| < \varepsilon$ for *any* points t and τ in $[a, b]$ such that $|t - \tau| < \eta$.

1.3.3 Most of the standard functions encountered in elementary calculus are defined by a single formula in each case, valid for all values of t, and are continuous everywhere. In particular this is so for the following functions:

(a) all polynomials; for example, $f(t) = at^2 + bt + c$,

(b) the exponential function e^t, and the hyperbolic functions $\sinh(t)$ and $\cosh(t)$,

(c) the trigonometric functions $\sin(t)$ and $\cos(t)$.

Any point at which a function $f(t)$ fails to be continuous is said to be a **discontinuity** of the function. In elementary calculus nothing worse than simple infinities are usually encountered and these are easily recognised (and easily avoided). For example, consider the function $f(t) = 1/t$ which is defined and continuous everywhere except at the point $t = 0$. As t approaches 0 from the right-hand side the function becomes indefinitely large and positive: as t approaches 0 from the left, the function becomes indefinitely large and negative. Behaviour of this kind, in which a function becomes unbounded in absolute value in the neighbourhood of some specific point, is typical of the discontinuities associated with the **rational** functions. A rational function f is a function which can always be expressed as the ratio of two polynomials:

$$f(t) = \frac{P(t)}{Q(t)} = \frac{a_n t^n + a_{n-1} t^{n-1} + \ldots + a_0}{b_m t^m + b_{m-1} t^{m-1} + \ldots + b_0},$$

and discontinuities of f will exist precisely at those points $t = t_k$ for which the denominator $Q(t)$ vanishes.

1.3.4 In many applications however, it is useful to consider functions which are bounded but admit sudden, discontinuous jumps in value, and to extend the operations and processes of the calculus so as to apply to them. It is precisely when we try to do this that the need for some generalisation of the concept of function first becomes apparent. This point will be taken up and dealt with in some detail in Chapter 2.

1.4 DIFFERENTIABILITY

1.4.1 Let the function f be defined on the open interval (a,b). The classical **first derivative** of f is the function f' defined by

$$\lim_{h \to 0} \left[\frac{f(t+h) - f(t)}{h} \right] \equiv f'(t) \equiv \frac{dy}{dt}$$

at all points t of (a,b) for which this limit exists. Note that h is not restricted to positive values, so that the limit is required to be uniquely defined no matter how the point $(t+h)$ approaches t: when this is the case the function f is said to be **differentiable** at the point t. Any function differentiable at a point t must certainly be continuous there. On the other hand f may be continuous at t and yet have no well-defined derivative there.

1.4.2 f is said to be **differentiable on the open interval** (a,b) if $f'(t)$ exists for every point t in (a,b). It is said to be **differentiable on the closed interval** $[a,b]$ if, in addition, the following limits exist

$$\lim_{h \downarrow 0} \left[\frac{f(a+h) - f(a)}{h} \right] \quad \text{and} \quad \lim_{h \downarrow 0} \left[\frac{f(b) - f(b-h)}{h} \right]$$

where, as indicated by the notation, h goes to 0 through positive values in each case.

If f is differentiable on (a,b) and if f' is continuous on this interval, then f is said to be **continuously differentiable**, or **smooth**, on (a,b). If both f and its first derivative f' are continuous on an interval, save possibly for a finite number of jump discontiuities, then f is said to be **sectionally smooth** on that interval.

A function f whose first derivative is itself differentiable on (a,b) is said to be **twice differentiable** on that interval; the first derivative of the function f' is said to be the **second derivative** of the function f, and is written as f''. If $y = f(t)$ then it is usual to write

$$f''(t) = \frac{d^2y}{dt^2}.$$

There are corresponding definitions for derivatives of higher order.

1.4.3 For differentiable functions in general the following results hold:

(i) If u and v are differentiable functions, and a and b are constants, then $w = au + bv$ is differentiable and

$$\frac{d}{dt}(au + bv) = a\left(\frac{du}{dt}\right) + b\left(\frac{dv}{dt}\right).$$

(ii) If u and v are differentiable then so also is the product function uv and

$$\frac{d}{dt}(uv) = v\frac{du}{dt} + u\frac{dv}{dt}.$$

Similarly

$$\frac{d}{dt}\left(\frac{u}{v}\right) = \frac{1}{v^2}\left\{v\frac{du}{dt} - u\frac{dv}{dt}\right\}.$$

(iii) If $y = f(x)$ where $x = g(t)$ then, provided the derivatives concerned do exist, we have

$$\frac{dy}{dt} = \frac{dy}{dx}\frac{dx}{dt}.$$

(iv) If $y = f(t)$ is a differentiable function with a well-defined inverse function, $t = f^{-1}(y)$, then

$$\frac{dy}{dt} = \left(\frac{dt}{dy}\right)^{-1}.$$

(v) **Rolle's Theorem.** Let f be continuous on the closed interval $[a, b]$ and differentiable at least in the open interval (a, b). If $f(a) = f(b) = 0$, then there exists a point t_0 in (a, b) such that $f'(t_0) = 0$.

1.5   TAYLOR'S THEOREM

1.5.1 The so-called **mean value theorems** of the differential calculus are more or less direct consequences of Rolle's theorem. In view of the extreme importance of these results, and of the consequences which can be derived from them, we give brief indications of how they may be established.

1.5.2 First Mean Value theorem. If f is a function which is continuous on the closed interval $[a, b]$ and differentiable at least on the open interval (a, b) then there exists a point t_0 in (a, b) such that

$$f(b) - f(a) = (b - a)f'(t_0).$$

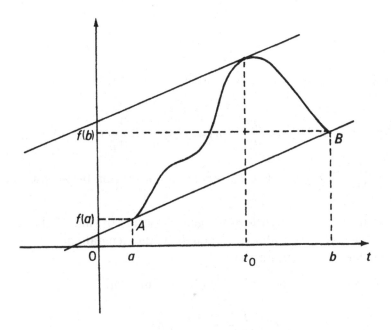

Figure 1.1:

Proof. The proof is almost obvious from Fig. 1.1.

The equation of the chord AB is given by

$$y = \left[\frac{f(b) - f(a)}{b - a}\right] t + \frac{bf(a) - af(b)}{b - a}$$

or,

$$y = \left[\frac{f(b) - f(a)}{b - a}\right](t - a) + f(a).$$

Let $F(t) \equiv f(t) - y$; then $F(b) = F(a) = 0$ and the result follows immediately on applying Rolle's Theorem to the function F.

1.5.3 Mean Value Theorem of Order n (Taylor's Theorem). If the function f and each of its first $(n-1)$ derivatives $f', f'', \ldots, f^{(n-1)}$, is continuous on the closed interval $[a, b]$, and if the nth derivative $f^{(n)}$ exists at least on the open interval (a, b), then there exists a point t_n in (a, b) such that

$$f(b) - f(a) = (b-a)f'(a) + \frac{(b-a)^2}{2!}f''(a) + \ldots$$

$$+ \frac{(b-a)^{n-1}}{(n-1)!}f^{(n-1)}(a) + \frac{(b-a)^n}{n!}f^{(n)}(t_n).$$

Alternatively, if we put $b = a + h$ then the result can be written in the form

$$f(a+h) = f(a) + hf'(a) + \frac{h^2}{2!}f''(a) + \ldots + \frac{h^{n-1}}{(n-1)!}f^{(n-1)}(a) + R_n$$

where $R_n = \frac{h^n}{n!}f^{(n)}(a + \theta_n h)$ and θ_n is some number lying between 0 and 1.

Proof. The proof is a straightforward generalisation of that of the first mean value theorem given above, although in the general case there is no simple geometric interpretation:

We have only to apply Rolle's Theorem to the function

$$F_n(t) - \left(\frac{b-t}{b-a}\right)^n F_n(a)$$

where

$$F_n(t) = f(b) - f(t) - (b-t)f'(t) - \frac{(b-t)^2}{2!}f''(t) - \ldots$$

$$- \frac{(b-t)^{n-1}}{(n-1)!}f^{(n-1)}(t).$$

1.5.4 Taylor's Series. Suppose now that the conditions for Taylor's theorem apply to a function f over an interval (which may be finite or infinite). Let a be a fixed point in that interval and write $t = a + h$ for any other point t within that interval. Then Taylor's Theorem states that

$$f(t) = f(a) + (t-a)f'(a) + \frac{(t-a)^2}{2!}f''(a) + \ldots$$

$$+ \frac{(t-a)^{n-1}}{(n-1)!}f^{(n-1)}(a) + R_n(t).$$

If, in addition, the following two conditions are fulfilled,

(i) f can be differentiated as often as we wish over the interval concerned (i.e. f is **infinitely differentiable** there),

(ii) for each t in the interval concerned we have

$$\lim_{n \to \infty} R_n(t) = 0,$$

then we can allow n to tend to infinity in Taylor's theorem and obtain the so-called **Taylor expansion of f about the point $t = a$**:

$$f(t) = f(a) + (t-a)f'(a) + \frac{(t-a)^2}{2!}f''(a) + \ldots$$

$$+ \frac{(t-a)^n}{n!}f^{(n)}(a) + \ldots$$

In the particular case $a = 0$ we get the **Maclaurin series** for f:

$$f(t) = f(0) + tf'(0) + \frac{t^2}{2!}f''(0) + \ldots + \frac{t^n}{n!}f^{(n)}(0) + \ldots$$

1.5.5 The reader is reminded that, among the standard elementary functions, the following expansions are particularly important:

$$e^t = 1 + t + \frac{t^2}{2!} + \frac{t^3}{3!} + \frac{t^4}{4!} + \ldots$$

$$\sinh(t) = t + \frac{t^3}{3!} + \frac{t^5}{5!} + \frac{t^7}{7!} + \ldots$$

$$\cosh(t) = 1 + \frac{t^2}{2!} + \frac{t^4}{4!} + \frac{t^6}{6!} + \ldots$$

$$\sin(t) = t - \frac{t^3}{3!} + \frac{t^5}{5!} - \frac{t^7}{7!} + \ldots$$

$$\cos(t) = 1 - \frac{t^2}{2!} + \frac{t^4}{4!} - \frac{t^6}{6!} + \dots$$

(Each of the above expansions is valid for all values of t.)

$$(1+t)^\alpha = 1 + \alpha t + \frac{\alpha(\alpha-1)}{2!}t^2 + \frac{\alpha(\alpha-1)(\alpha-2)}{3!}t^3 + \dots$$

(If α is a positive integer the series terminates and reduces to the ordinary binomial expansion. In every other case the expansion is an infinite series valid for all values of t such that $-1 < t < +1$.)

$$\log(1+t) = t - \frac{t^2}{2} + \frac{t^3}{3} - \frac{t^4}{4} + \dots, \qquad -1 < t \le +1,$$

$$\log(1-t) = -t - \frac{t^2}{2} - \frac{t^3}{3} - \frac{t^4}{4} - \dots, \qquad -1 \le t < +1,$$

$$\log\left[\frac{1+t}{1-t}\right] = 2\left[t + \frac{t^3}{3} + \frac{t^5}{5} + \frac{t^7}{7} + \dots\right], \qquad -1 < t < +1,$$

$$\arctan(t) = t - \frac{t^3}{3} + \frac{t^5}{5} - \frac{t^7}{7} + \dots, \qquad -1 \le t \le +1.$$

1.5.6 l'Hopital's Rule.

In order to evaluate limits of the form $\lim_{t\to t_0}[f(t)/g(t)]$ where we have

$$\lim_{t\to t_0} f(t) = \lim_{t\to t_0} g(t) = 0$$

we can often make use of a simple corollary of Taylor's theorem known as **l'Hopital's Rule**. Suppose that f and g have continuous derivatives in some neighbourhood of the point t_0 and that $f(t_0) = g(t_0) = 0$ while $g'(t_0) \ne 0$. By Taylor's theorem (with $n = 1$)

$$\frac{f(t_0+h)}{g(t_0+h)} = \frac{f(t_0) + hf'(t_0+\theta h)}{g(t_0) + hg'(t_0+\theta' h)} = \frac{f'(t_0+\theta h)}{g'(t_0+\theta' h)},$$

and so

$$\lim_{t\to t_0} \frac{f(t)}{g(t)} = \lim_{h\to 0} \frac{f(t_0+h)}{g(t_0+h)} = \lim_{h\to 0} \frac{f'(t_0+\theta h)}{g'(t_0+\theta' h)} = \frac{f'(t_0)}{g'(t_0)}.$$

In fact a stronger and more useful form of the rule can be established: if $f(t) \to 0$ and $g(t) \to 0$ as $t \to t_0$ then

$$\lim_{t\to t_0} \frac{f(t)}{g(t)} = \lim_{t\to t_0} \frac{f'(t)}{g'(t)},$$

provided the latter limit exists. In this form it is not necessary to assume that $f(t_0)$ and $g(t_0)$ are actually defined.

1.6 INTEGRATION

1.6.1 Let f be a function of the real variable t which is continuous for all t in the closed interval $[a, b]$. Subdivide this interval into n sub-intervals by arbitrarily chosen points $t_1, t_2, \ldots, t_{n-1}$, and write $t_0 = a$ and $t_n = b$. In each sub-interval choose some point ξ_k arbitrarily and form the sum

$$f(\xi_1)(t_1 - a) + f(\xi_2)(t_2 - t_1) + \ldots + f(\xi_n)(b - t_{n-1}) \equiv \sum_{k=1}^{n} f(\xi_k)\Delta_k$$

Now let the number of subdivisions increase in such a way that, for each k, the number $\Delta_k \equiv t_k - t_{k-1}$ tends to zero. Then it can be shown that the sums tend to a unique limit (independently of how the points of subdivision t_k, and the points ξ_k in the resulting sub-intervals, have been chosen). This limit is, by definition, the elementary, or **Riemann**, integral of the continuous function f from $t = a$ to $t = b$. More precisely, given any number $\varepsilon > 0$, however small, we can always find a positive number h such that

$$\left| \int_a^b f(t)dt - \sum_{k=1}^{n} f(\xi_k)\Delta_k \right| < \varepsilon$$

for any sum of the form described above in which $\Delta_k < h$ for every k.

1.6.2 As is well known, the geometric significance of the integral lies in the fact that it gives a measure of the area enclosed by the graph of the function $y = f(t)$, the ordinates $t = a$ and $t = b$, and the t-axis (with the convention that areas lying above the axis are taken to be positive while those lying below are taken to be negative). The other important aspect of integration is its role as, in some sense, a process inverse to differentiation. This connection follows as a result of the **first mean value theorem** of the integral calculus:

First Mean Value Theorem (for integrals). If f is continuous on the closed interval $[a, b]$ then there must exist exist some point ξ in (a, b) such that

$$(b - a)f(\xi) = \int_a^b f(t)dt.$$

Proof. As in the case of the first mean value theorem of the differential calculus the geometrical significance of the theorem (Fig. 1.2) is almost obvious.

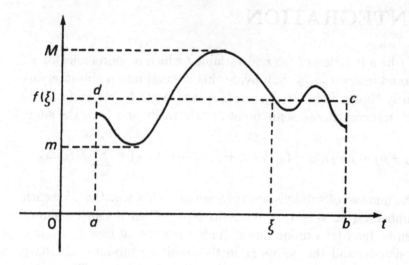

Figure 1.2: Area of rectangle $abcd = (b - a)f(\xi)$.

Let m and M denote respectively the least and greatest of the values which the continuous function f assumes in the interval $[a, b]$. Then for any sub-division by points t_k we have

$$m\Delta_k \le f(\xi_k)(t_k - t_{k-1}) \le M\Delta_k.$$

Hence,

$$m\sum_{k=1}^{n}(t_k - t_{k-1}) \le \sum_{k=1}^{n} f(\xi_k)(t_k - t_{k-1}) \le M\sum_{k=1}^{n}(t_k - t_{k-1})$$

so that, in the limit,

$$m(b - a) \le \int_a^b f(t)dt \le M(b - a).$$

Since f is continuous on $[a, b]$ it will take on every value between m and M, its least and greatest values there. Hence there must exist some point ξ in (a, b) such that

$$\frac{1}{b - a}\int_a^b f(t)dt = f(\xi).$$

A simple extension of this argument yields the following generalisation of the integral mean value theorem:

Generalised Mean Value Theorem. If f is continuous on $[a, b]$, and if g is any positive continuous function on $[a, b]$, then there exists some point ξ in (a, b) such that

$$\int_a^b f(t)g(t)dt = f(\xi) \int_a^b g(t)dt.$$

1.6.3 Fundamental Theorem of the Calculus. If f is continuous on $[a, b]$, define a function F on $[a, b]$ by writing

$$F(t) = \int_a^t f(\tau)d\tau.$$

Then,

$$\frac{F(t+h) - F(t)}{h} = \frac{1}{h}\left[\int_a^{t+h} f(\tau)d\tau - \int_a^t f(\tau)d\tau\right]$$

$$= \frac{1}{h}\int_t^{t+h} f(\tau)d\tau = f(\xi)$$

for some ξ in $(t, t + h)$. If we allow h to tend to zero then, since f is continuous, $f(\xi)$ tends to $f(t)$ as its limit, and so

$$F'(t) = \lim_{h \to 0} \frac{F(t+h) - F(t)}{h} = f(t).$$

Now let F denote any function whose derivative is f. Then we can always write

$$F(t) = \int_a^t f(\tau)d\tau + C$$

where C is a constant. Putting $t = a$ in this gives

$$F(a) = \int_a^a f(\tau)d\tau + C = C.$$

Hence, putting $t = b$, we obtain the result usually referred to as the **Fundamental Theorem of the Calculus:**

$$\int_a^b f(\tau)d\tau = F(b) - F(a)$$

where F is any function whose derivative is equal to the integrand f on $[a, b]$. Such a function F is often called a **primitive** or an **anti-derivative** of f on $[a, b]$.

Integration by parts. If f is continuous on $[a, b]$ and F is a primitive of f, and if g is continuously differentiable on $[a, b]$, then

$$\int_a^b f(t)g(t)dt = [F(b)g(b) - F(a)g(a)] - \int_a^b F(t)g'(t)dt.$$

This formula is an immediate consequence of the rule for the derivative of a product, and is well known in elementary calculus as a device for transforming certain integrals into forms which can be evaluated more easily. As will be seen hereafter, it can take on a wider significance if the conditions on f and g are modified so that it can form the basis of a generalisation of the concept of derivative itself.

1.6.4 Remark. The definition of the integral as the limit of finite sums is of course not confined to continuous functions. If f is defined and bounded on $[a, b]$ then we can always construct approximating **Riemann sums**

$$\sum_{k=1}^n f(\xi_k)\Delta_k$$

but without the hypothesis of continuity it does not necessarily follow that these sums converge to a definite limit as the sub-divisions are taken smaller and smaller. Whenever such a limit does exist the function f, whether it is continuous or not, is said to be **integrable in the Riemann sense**, (integrable-R), over $[a, b]$. This will be the case, for example, when f has nothing worse than finitely many jump discontinuities on $[a, b]$. Note that whenever f is Riemann-integrable over $[a, b]$ we can always define the function

$$F(t) = \int_a^t f(\tau)d\tau.$$

Further,

$$|F(t+h) - F(t)| = \left| \int_t^{t+h} f(\tau)d\tau \right| < |h| \sup_{t \le \tau \le t+h} |f(\tau)|.$$

Since f is bounded on $[a, b]$ it follows that $|F(t+h) - F(t)|$ tends to zero with h, i.e. the function F is continuous on $[a, b]$ whether the integrand f is itself continuous or not. (Integration is a smoothing operation.)

1.7 IMPROPER INTEGRALS

1.7.1 Suppose that f is bounded and continuous (more generally, bounded and Riemann-integrable) over every interval $[a, t]$, where a is some fixed number and t is any number greater than a. The **infinite** or **improper** Riemann integral $\int_a^{+\infty} f(t)dt$ is defined by the relation

$$\int_a^{+\infty} f(t)dt \equiv \lim_{t \to \infty} \int_a^t f(\tau)d\tau = \lim_{t \to \infty} F(t) - F(a)$$

where F denotes any primitive of f. If the limit exists as a well-defined finite number then the integral is said to **converge**; otherwise it is said to **diverge**. An integral of this type, in which the range of integration becomes infinite but the integrand remains bounded, is called an **improper integral of the first kind**.

If $\int_a^{+\infty} |f(t)|dt$ converges then it can be shown that $\int_a^{+\infty} f(t)dt$ must also converge, and in this case the convergence is said to be **absolute**. If $\int_a^{+\infty} |f(t)|dt$ diverges then the integral $\int_a^{+\infty} f(t)dt$ may diverge or it may converge: in the latter event the convergence is said to be **conditional**.

A corresponding treatment applies for the definition

$$\int_{-\infty}^{a} f(t)dt \equiv \lim_{t \to \infty} \int_{-t}^{a} f(\tau)d\tau = F(a) - \lim_{t \to \infty} F(-t).$$

Finally, to define an improper integral of the first kind over the whole range $(-\infty, +\infty)$, we write

$$\int_{-\infty}^{+\infty} f(t)dt \equiv \lim_{t_1 \to \infty} \int_{-t_1}^{a} f(t)dt + \lim_{t_2 \to \infty} \int_{a}^{t_2} f(t)dt.$$

Comparison Test (integrals of the first kind). Let f and g be bounded integrable functions of t for $a \leq t \leq x$, and suppose that $0 \leq f(t) \leq g(t)$ throughout this range. Then

(i) if $\int_a^{+\infty} g(t)dt$ converges, then so also does $\int_a^{+\infty} f(t)dt$,

(ii) if $\int_a^{+\infty} f(t)dt$ diverges, then so also does $\int_a^{+\infty} g(t)dt$.

A particularly useful application of this general test is obtained by taking $g(t) = t^{-p}$. For, if $a > 0$ and $p \neq 1$, then we get

$$\int_a^{+\infty} \frac{dt}{t^p} = \lim_{T \to \infty} \int_a^T \frac{dt}{t^p} = \lim_{T \to \infty} \left[\frac{t^{1-p}}{1-p} \right]_a^T.$$

This converges to the value $a^{1-p}/(p-1)$ if $p > 1$ and diverges to $+\infty$ if $p < 1$. In case $p = 1$ we have

$$\int_a^{+\infty} \frac{dt}{t} = \lim_{T\to\infty} \int_a^T \frac{dt}{t} = \lim_{T\to\infty} [\log(T) - \log(a)]$$

and this diverges to $+\infty$.

Accordingly we derive the simple and important 'p-test' for improper integrals of the first kind:

Let f be a bounded continuous function which is non-negative for all $t \geq a$ (where $a > 0$). If there exists a number p such that $\lim_{t\to a} t^p f(t) = A$ then

(i) $\int_a^{+\infty} f(t)dt$ converges if $p > 1$ and A is finite,

(ii) $\int_a^{+\infty} f(t)dt$ diverges if $p \leq 1$ and $A > 0$ (possibly infinite).

1.7.2 Now suppose that f is a function which becomes unbounded as t approaches a in the interval $[a, b]$. We define

$$\int_a^b f(t)dt \equiv \lim_{\varepsilon\downarrow 0} \int_{a+\varepsilon}^b f(t)dt$$

whenever this limit exists. Similarly if f becomes unbounded as t tends to b in $[a, b]$ then we define

$$\int_a^b f(t)dt \equiv \lim_{\varepsilon\downarrow 0} \int_a^{b-\varepsilon} f(t)dt.$$

Finally, if t_0 is some point in (a, b) and if f becomes unbounded in the neighbourhood of t_0 then the improper integral $\int_a^b f(t)dt$ is said to converge to the value

$$\lim_{\varepsilon_1\downarrow 0} \int_a^{t_0-\varepsilon_1} f(t)dt + \lim_{\varepsilon_2\downarrow 0} \int_{t_0+\varepsilon_2}^b f(t)dt$$

provided that the limits exist independently (that is, ε_1 and ε_2 must be allowed to tend to zero independently of one another and we require that both the integrals

$$\int_a^{t_0} f(t)dt \text{ and } \int_{t_0}^b f(t)dt$$

should converge).

Integrals such as these, in which the range of integration is finite but the integrand becomes unbounded at one or more points of that range, are called **improper integrals of the second kind**. When we have both an infinite range of integration and an integrand which becomes unbounded within that range, we speak of an **improper integral of the third kind**.

Comparison Test (integrals of the second kind). Let f and g be continuous for $a < t \leq b$, and suppose that $0 \leq f(t) \leq g(t)$ throughout this range. Suppose also that both f and g become unbounded as t tends to a. Then

(i) if $\int_a^b g(t)dt$ converges, then so also does $\int_a^b f(t)dt$,

(ii) if $\int_a^b f(t)dt$ diverges, then so also does $\int_a^b g(t)dt$.

Once again we obtain a convergence test of particular importance by taking $g(t) = t^{-p}$. If $p \neq 1$ then

$$\int_0^1 \frac{dt}{t^p} = \lim_{\varepsilon \downarrow 0} \int_\varepsilon^1 \frac{dt}{t^p} = \frac{1}{1-p} - \lim_{\varepsilon \downarrow 0}\left[\frac{\varepsilon^{1-p}}{1-p}\right]$$

which converges to the value $1/(1-p)$ if $p < 1$ and diverges to $+\infty$ if $p > 1$. Also if $p = 1$ we have

$$\int_0^1 \frac{dt}{t} = \lim_{\varepsilon \downarrow 0} \int_\varepsilon^1 \frac{dt}{t} = \log(1) - \lim_{\varepsilon \downarrow 0}[\log(\varepsilon)]$$

and this diverges to $+\infty$. As a result we obtain a corresponding 'p-test' for improper integrals of the second kind:

Let f be a continuous, non-negative function for $a < t \leq b$ and suppose that $f(t)$ becomes unbounded as t tends to a. If there exists a number p such that $\lim_{t \to a}(t-a)^p f(t) = A$ then

(i) $\int_a^b f(t)dt$ converges if $p < 1$ and A is finite,

(ii) $\int_a^b f(t)dt$ diverges if $p \geq 1$ and $A > 0$ (possibly infinite).

The integral sign may sometimes be used in a purely conventional sense, as shown in tne next section for example.

1.7.3 Cauchy Principal Value. Suppose that f is a function which becomes unbounded in the neighbourhood of a point t_0 in (a, b) and that the improper integral $\int_a^b f(t)dt$ diverges. It may happen that although the limits

$$\lim_{\varepsilon_1 \downarrow 0} \int_a^{t_0 - \varepsilon_1} f(t)dt \text{ and } \lim_{\varepsilon_2 \downarrow 0} \int_{t_0 + \varepsilon_2}^b f(t)dt$$

do not exist independently, we may set $\varepsilon_1 = \varepsilon_2 = \varepsilon$ and then find that the unbounded parts of the two integrals cancel each other out. This gives a certain finite answer called the **Cauchy principal value** of the (divergent) integral. We usually write

$$P \int_a^b f(t)dt \equiv \lim_{\varepsilon \downarrow 0} \left[\int_a^{t_0 - \varepsilon} f(t)dt + \int_{t_0 + \varepsilon}^b f(t)dt \right].$$

For example, $P \int_{-1}^{+1} \frac{dt}{t} = 0$, although each of the integrals $\int_{-1}^{-\varepsilon} \frac{dt}{t}$ and $\int_{\varepsilon}^1 \frac{dt}{t}$ become arbitrarily large in absolute value as ε tends to 0.

1.8 UNIFORM CONVERGENCE

1.8.1 A sequence (f_m) of functions converges to a limit function f in the simple or **pointwise** sense if

$$\lim_{m \to \infty} f_m(t) = f(t)$$

for each point t in the domain of definition concerned. This means that given any number $\varepsilon > 0$ we can find, for each particular value of t, a corresponding integer $m_0 = m_0(\varepsilon, t)$ such that

$$|f(t) - f_m(t)| < \varepsilon$$

for all $m \geq m_0$. As our notation indicates, the number m_0 will generally depend on the particular value of t concerned as well as on the number ε. Now suppose that (f_m) is a sequence of (bounded) functions which converges to a limit function f not just in the simple pointwise sense but in such a way that the following additional condition holds:

$$\lim_{m \to \infty} \left[\sup_t |f(t) - f_m(t)| \right] = 0.$$

This means that given any $\varepsilon > 0$ we can always find a corresponding integer $m_0 = m_0(\varepsilon)$ such that for all $m \geq m_0$ we have

$$\sup_t |f(t) - f_m(t)| < \varepsilon.$$

It follows that

$$|f(t) - f_m(t)| < \varepsilon$$

for all $m \geq m_0$ and for *every* t. In this case the integer m_0 depends only on ε and is independent of t. The convergence of the f_m to f is then said to be **uniform**. Uniform convergence is a much stronger form of convergence than the ordinary pointwise sense, and a number of important properties of uniformly convergent sequences can be derived.

1.8.2 Theorem. If (f_m) is a sequence of bounded, continuous functions converging uniformly to f then f is bounded and continuous.

Proof. For any t and x we have

$$|f(t+x) - f(t)| \leq |f(t+x) - f_m(t+x)|$$
$$+ |f_m(t+x) - f_m(t)| + |f_m(t) - f(t)|.$$

Given $\varepsilon > 0$ we can find a corresponding positive integer $m_0 = m_0(\varepsilon)$ such that for all t and x

$$|f(t+x) - f_{m_0}(t+x)| < \frac{\varepsilon}{3} \text{ and } |f_{m_0}(t) - f(t)| < \frac{\varepsilon}{3}.$$

Again, by the continuity of f_{m_0} at t, we can always find $\eta > 0$ such that

$$|f_{m_0}(t+x) - f_{m_0}(t)| < \frac{\varepsilon}{3} \text{ for all } x \text{ such that } |x| < \eta.$$

It follows at once that f is continuous. That it is also bounded is a consequence of the fact that for all t

$$f_{m_0}(t) - \varepsilon < f(t) < f_{m_0}(t) + \varepsilon.$$

1.8.3 Theorem. If f is the uniform limit of a sequence (f_m) of bounded, continuous functions on the finite, closed interval $[a, b]$ then

$$\int_a^b f(t)dt = \lim_{m \to \infty} \int_a^b f_m(t)dt.$$

Proof. If the sequence (f_m) converges uniformly to f then by the Theorem of 1.8.2 the limit function f is certainly bounded and continuous, and is therefore integrable over $[a, b]$.
Also, for each $m \in \mathbb{N}$ we have,

$$\left| \int_a^b f(t)dt - \int_a^b f_m(t)dt \right| \le \int_a^b |f(t) - f_m(t)|dt$$

$$\le (b-a)\sup_t |f(t) - f_m(t)|$$

and the result follows since $\lim_{m\to\infty} \{\sup|f(t) - f_m(t)|\} = 0$.

Corollary. Suppose further that each of the derivatives f'_m exists and is continuous on $[a, b]$ and that the sequence (f'_m) converges uniformly to a function ϕ on $[a, b]$. Then $\phi(t) = f'(t)$.

Proof. The uniformity of the convergence shows that ϕ must be continuous on $[a, b]$. For any t in $[a, b]$ we have

$$\int_a^t \phi(\tau)d\tau = \lim_{m\to\infty} \int_a^t f'_m(\tau)d\tau = \lim_{m\to\infty} [f_m(t) - f_m(a)]$$

$$= f(t) - f(a).$$

The Fundamental Theorem of the Calculus then shows that f is differentiable on $[a, b]$ and that $f'(t) = \phi(t)$.

1.9   DIFFERENTIATING INTEGRALS

1.9.1 From the Fundamental Theorem of the Calculus we know that if f is a function bounded and continuous on the finite closed interval $[a, b]$ then

$$\int_a^b f(t)dt = F(b) - F(a),$$

where f is a function such that $F'(t) = f(t)$ on $[a, b]$. Hence if either the top limit b or the bottom limit a is allowed to vary then it follows that

$$\frac{d}{db} \left[\int_a^b f(t)dt \right] = \frac{d}{db}[F(b) - F(a)] = f(b),$$

and

$$\frac{d}{da} \left[\int_a^b f(t)dt \right] = \frac{d}{da}[F(b) - F(a)] = -f(a).$$

1.9.2 Now let $f(x, y)$ be a function defined and continuous throughout some region R of the (x, y)-plane and such that both the partial derivatives $\partial f / \partial x$ and $\partial f / \partial y$ exist and are continuous at each point of $\mathbb{R}$.

Let the functions $a = \phi_1(x)$ and $b = \phi_2(x)$ have continuous derivatives throughout the interval $a \le x \le b$, and consider the function

$$G(x) \equiv \int_a^b f(x,y)dy \equiv G(a,b,x).$$

By the chain rule for partial derivatives we have

$$\frac{dG}{dx} = \frac{\partial G}{\partial x} + \frac{\partial G}{\partial a}\frac{da}{dx} + \frac{\partial G}{\partial b}\frac{db}{dx}.$$

Then

(i) $\dfrac{\partial G}{\partial a} = \dfrac{\partial}{\partial a}\displaystyle\int_a^b f(x,y)dy = -f(x,a)$

(ii) $\dfrac{\partial G}{\partial b} = \dfrac{\partial}{\partial b}\displaystyle\int_a^b f(x,y)dy = f(x,b)$

(iii) $\dfrac{\partial G}{\partial x} \equiv \lim\limits_{h\to 0}\dfrac{1}{h}\left[\displaystyle\int_a^b f(x+h,y)dy - \int_a^b f(x,y)dy\right]$

$$= \lim_{h\to 0}\int_a^b \frac{1}{h}[f(x+h,y)-f(x,y)]dy$$

$$= \lim_{h\to 0}\int_a^b f_x(x+\theta h,y)dy$$

where $0 < \theta < 1$. Since $f_x(x,y) \equiv \partial f/\partial x$ is assumed to be continuous in $[a,b]$ we can write

$$f_x(x+\theta h, y) = f_x(x,y) + \varepsilon,$$

where $|\varepsilon|$ tends to 0 with h. It follows that

$$\frac{\partial G}{\partial x} = \int_a^b \frac{\partial}{\partial x}f(x,y)dy.$$

Thus finally we obtain what is known as **Leibnitz's Rule** for differentiating an integral with respect to a parameter:

$$\frac{d}{dx}\int_{\phi_1(x)}^{\phi_2(x)} f(x,y)dy = \int_{\phi_1(x)}^{\phi_2(x)} \frac{\partial}{\partial x}f(x,y)dy$$

$$-\phi_1'(x)f(x,\phi_1(x)) + \phi_2'(x)f(x,\phi_2(x)).$$

Chapter 2

The Dirac Delta Function

2.1 THE UNIT STEP FUNCTION

2.1.1 Simple discontinuities. The functions we meet in elementary calculus are usually expected to be continuous throughout the range of values with which we happen to be concerned. But it eventually becomes necessary to deal with situations in which discontinuities may occur and may be of essential significance. Consider therefore, a point t_0 within the domain of definition of a function f at which continuity may fail. Any point t to the right of t_0 can be expressed as $t = t_0 + h$, where $h > 0$. As h tends to zero the point $(t_0 + h)$ approaches t_0 from the right. If the corresponding values $f(t_0 + h)$ tend to a limiting value then this value is called **the right-hand limit** of f as t tends to t_0 and is usually written as $f(t_0+)$:

$$f(t_0+) = \lim_{h \downarrow 0} f(t_0 + h), \quad h > 0. \tag{2.1}$$

If, in addition, this limit is equal to the value $f(t_0)$ which the function f actually assumes at the point t_0 then f is said to be **continuous from the right** at t_0.

Similarly any point to the left of t_0 can be expressed as $t = t_0 - h$ where again h is positive. The limit, if it exists, of $f(t_0 - h)$ as h tends to 0 through positive values is called the **left-hand limit** of f as t tends to t_0 and is usually written as $f(t_0-)$. If this limit does exist, and is equal to the value $f(t_0)$, then f is said to be **continuous from the left** at t_0.

Clearly f is continuous at t_0 in the usual sense if and only if it is both continuous from the right and continuous from the left there:

$$\lim_{h\downarrow,0} f(t_0 + h) = \lim_{h\downarrow 0} f(t_0 - h) = f(t_0). \qquad (2.2)$$

If f is continuous at a point t_0 then it must certainly be the case that $f(t_0)$ is defined. A comparatively trivial example of a discontinuity is afforded by a formula like

$$\frac{\sin(t)}{t} \qquad (2.3)$$

which defines a function continuous everywhere except at the point $t = 0$ where the formula becomes meaningless. However we have

$$f(0+) = \lim_{h\downarrow 0} \frac{\sin(h)}{h} = 1 \quad \text{and} \quad f(0-) = \lim_{h\downarrow 0} \frac{\sin(-h)}{(-h)} = 1 \qquad (2.4)$$

and so the discontinuity may easily be removed by defining $f(0)$ to be 1. In general, a function f is said to have a **removable discontinuity** at t_0 if both the right-hand and the left-hand limits of f at t_0 exist and are equal, but $f(t_0)$ is either undefined or else has a value different from $f(t_0+)$ and $f(t_0-)$. The discontinuity disappears on suitably defining (or re-defining) $f(t_0)$. If the one-sided limits $f(t_0+)$ and $f(t_0-)$ both exist but are unequal in value then f is said to have a **simple**, or **jump**, discontinuity at t_0. The number $f(t_0+)-f(t_0-)$ is then called the **saltus**, or the **jump**, of the function at t_0. The simplest example of such a discontinuity occurs in the definition of a special function which is of particular importance.

2.1.2 Step functions. We shall define the **Heaviside unit step function**, u, as that function which is equal to 1 for every positive value of t and equal to 0 for every negative value of t. This function could equally well have been defined in terms of a specific formula; for example

$$u(t) = \frac{1}{2}\left[1 + \frac{t}{|t|}\right]. \qquad (2.5)$$

Figure 2.1 shows the graph of $y = u(t)$, together with those of functions $u(t - a)$ and $u(a - t)$ obtained by translation and reflection.

The value $u(0)$ is left undefined here. In many contexts, as will become clear later, this does not matter. Nevertheless, some texts do specify a particular value for $u(0)$, the most popular candidates being $u(0) = 0$, $u(0) = 1$ and $u(0) = \frac{1}{2}$. Some plausible justification could be given for each one of these alternative choices. Where necessary we shall circumvent this problem by adopting the following convention:

The symbol u will always refer to the unit step function, defined as above, with $u(0)$ left unspecified. For any given real number c we shall understand by u_c the function obtained when we complete the definition by assigning the value c at the origin. Thus we have,

$$u_c(t) = u(t) \quad \text{for all } t \neq 0 \quad \text{and} \quad u_c(0) = c. \tag{2.6}$$

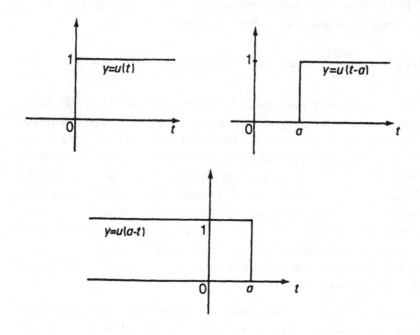

Figure 2.1:

Note that taking $c = 0$ gives a function u_0 which is continuous from the left at the origin: similarly, taking $c = 1$ gives a function u_1 which is continuous from the right there.

The Heaviside unit step function is of particular importance in the context of control theory, electrical network theory and signal processing. However its immediate general significance can be gauged from the following considerations: suppose that ϕ is a function which is continuous everywhere except for the point $t = a$, at which it has a simple discontinuity (Fig. 2.2). Then ϕ can always be represented as a linear combination of functions ϕ_1 and ϕ_2 which are continuous everywhere but which have been truncated at $t = a$:

$$\phi(t) = \phi_1(t)u(a - t) + \phi_2(t)u(t - a) \qquad (2.7)$$

Then $\phi(t) = \phi_1(t)$ for all $t < a$ and $= \phi_2(t)$ for all $t > a$.

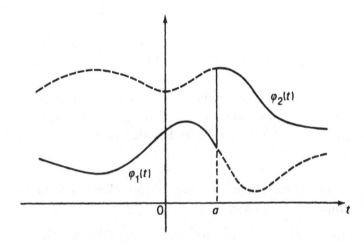

Figure 2.2:

In particular, note the linear combination of unit step functions which produces the so-called **signum** function:

$$\mathrm{sgn}(t) = u(t) - u(-t). \qquad (2.8)$$

The signum function takes the value $+1$ for all positive values of t, the value -1 for all negative values of t, and is left undefined at the origin. Once again usage differs in this respect, and a number of authorities assign the value 0 to $\mathrm{sgn}(0)$: where necessary we shall denote by $\mathrm{sgn}_0(t)$ the function obtained by completing the definition of $\mathrm{sgn}(t)$ in this manner.

2.1.3 Sectionally continuous functions. Any function which is continuous in an interval except for a finite number of simple discontinuities is said to be **sectionally continuous** or **piecewise continuous** on that interval. In particular suppose that a finite interval $[a, b]$ is sub-divided by points $t_0, t_1, \ldots, t_n$ where

$$a = t_0 \le t_1 \le t_2 \le \ldots \le t_n = b.$$

If s is a function which is constant on each of the (open) sub-intervals (t_{k-1}, t_k), for example, if

$$s(t) = \alpha_k, \quad \text{for } t_{k-1} < t < t_k \quad (k = 1, 2, ..., n),$$

then s is certainly piecewise continuous on $[a, b]$. Such a function is often called a **step-function** or a **simple function** on $[a, b]$. The particular function $u(t)$ is called the Heaviside unit step in honour of the English electrical engineer and applied mathematician **Oliver Heaviside**, and is often denoted by the alternative symbol, $H(t)$. Other notations such as $U(t), \sigma(t)$ and $\eta(t)$ are also to be found in the literature.

Exercises I.

1. Identify the functions defined by each of the following formulas and draw sketch graphs in each case:

(a) $u[(t - a)(t - b)]$; (b) $u(e^t - \pi)$; (c) $u(t - \log \pi)$;

(d) $u(\sin t)$; (e) $u(\cos t)$; (f) $u(\sinh t)$; (g) $u(\cosh t)$.

2. Identify the functions defined by each of the following formulas and draw sketch graphs in each case:

(a) $\text{sgn}(t^2 - 1)$; (b) $\text{sgn}(e^{-t})$; (c) $\text{sgn}(\tan t)$;

(d) $\text{sgn}\left(\sin\left(\dfrac{1}{t}\right)\right)$; (e) $t^2 \text{sgn}(t)$; (f) $\sin(t) . \text{sgn}(\sin t)$;

(g) $\sin(t) . \text{sgn}(\cos(t))$.

3. Find each of the following limit functions:

(a) $\lim_{n \to \infty} \left[\frac{1}{2} + \frac{1}{\pi} \arctan(nt)\right]$,

where $\arctan(t)$ is understood to denote the principal value of the inverse tangent, viz.: $-\pi/2 < \arctan(t) \le \pi/2$.

(b) $\lim_{n \to \infty} [\exp(-e^{-nt})]$.

4. A whole family of useful and important functions can be generated by replacing the variable t by $|t|$ in the familiar formulas for functions encountered in elementary calculus. The functions thus obtained are generally differentiable except at isolated points where the derivatives have jump discontinuities. Find the derivatives of each of the functions listed below and sketch the graphs of the function and its derivative in each case.

(a) $1 - |t|^3$; (b) $e^{|t|}$; (c) $e^{-|t|}$; (d) $\sin(|t|)$;

(e) $|\sin(t)|$; (f) $|\sin(|t|)|$; (g) $\sinh(|t|)$.

5. The examples of the preceding question should make it clear that continuity does not always imply differentiability. Although the reverse implication is in fact valid, it is instructive to realise that the mere existence of a derivative at every point is no guarantee of 'reasonable' behaviour:

(a) Prove that if f is differentiable at a point t_0 then it must necessarily be continuous there.

(b) Let $f(t) = t^2 \sin(1/t)$ for $t \neq 0$, and $f(0) = 0$. Show that f is differentiable everywhere but that the derivative f' is neither continuous from the right nor continuous from the left at the origin.

(c) Let $g(t) = t^2 \sin(1/t^2)$ for $t \neq 0$, and $g(0) = 0$. Show that g is differentiable everywhere, but that its derivative g' is not even bounded in the neighbourhood of the origin.

6. A function g is said to be **absolutely continuous** if the following criterion is satisfied: there exists a function h which is integrable over every finite interval $[a, b]$ within the domain of definition of g and which is such that

$$g(b) - g(a) = \int_a^b h(t)dt.$$

Give an example to show that an absolutely continuous function need not be differentiable everywhere. Are the functions defined in question (4) above absolutely continuous?

2.2 DERIVATIVE OF THE UNIT STEP FUNCTION

2.2.1 Naive definition of delta function. Suppose that we try to extend the definition of differentiation in such a way that it applies to functions with jump discontinuities. In particular we would need to define a 'derivative' for the unit step function, u. For all $t \neq 0$ this is of course well-defined in the classical sense as:

$$u'(t) = 0 \quad \text{for all} \ t \neq 0,$$

corresponding to the obvious fact that the graph of $y = u(t)$ has zero slope for all non-zero values of t. At $t = 0$, however, there is a jump discontinuity, and the definition of derivative accordingly fails. A glance at the graph suggests that it would not be unreasonable to describe the slope as 'infinite' at this point. Moreover, if we take any specific representation, u_c, of the unit step, then the ratio $\frac{u_c(h)-u_c(0)}{h}$ becomes arbitrarily large as h approaches 0, or using the familiar convention,

$$\lim_{h \to 0} \frac{u_c(h) - u_c(0)}{h} = +\infty.$$

Thus, from a descriptive point of view at least, the derivative of u would appear to be a function which (for the moment) we shall denote by $\delta(t)$ and which has the following pointwise specification:

$$\delta(t) \equiv u'(t) = 0, \quad \text{for all} \ t \neq 0, \qquad (2.7a)$$

and, once again adopting the conventional use of the ∞ sign,

$$\delta(0) = +\infty. \qquad (2.7b)$$

However, as will shortly become clear, we shall need eventually to consider more carefully the significance of the symbol $\delta(t)$. Does it really denote a function, in the proper sense of the word, and in what sense does it represent the derivative of the function $u(t)$?

2.2.2 Now let f be any function continuous on a neighbourhood of the origin, say for $-a < t < +a$. Then we should have

$$\int_{-a}^{+a} f(t)u'(t)dt = \int_{-a}^{+a} f(t) \lim_{h \to 0} \left[\frac{u(t+h) - u(t)}{h} \right] dt.$$

Assuming that it is permissible to interchange the operations of integration and of taking the limit as h tends to 0, this gives

$$\int_{-a}^{+a} f(t)\delta(t)dt = \lim_{h \to 0} \frac{1}{h} \int_{-h}^{0} f(t)dt = \lim_{h \to 0} f(\xi),$$

where ξ is some point lying between $-h$ and 0 (using the First Mean Value Theorem of the Integral Calculus). Since f is assumed continuous in the neighbourhood of $t = 0$, it follows that

$$\int_{-a}^{+a} f(t)\delta(t)dt = f(0). \tag{2.8}$$

If f happens to be continuously differentiable in the neighbourhood of the origin then the same result can be obtained by a formal application of integration by parts (although the critical importance of the order in which the operations of integration and of proceeding to the limit are carried out is not made manifest). Thus we would appear to have

$$\int_{-a}^{+a} f(t)u'(t)dt = [u(t)f(t)]_{-a}^{+a} - \int_{-a}^{+a} f'(t)u(t)dt$$

which, since $u(t) = 0$ for $t < 0$, reduces to

$$f(a) - \int_{0}^{a} f'(t)dt = f(a) - [f(a) - f(0)] = f(0).$$

2.2.3 Sampling property of the delta function. The result (2.8) for arbitrary continuous functions f is usually referred to as the **sampling property** of δ. If we follow the usual conventional language, still to be found in much of the engineering and physics literature then we would refer to δ as the **delta function**, or more specifically as the **Dirac delta function**. This sampling property is clearly independent of the actual value of the number a, and depends only on the behaviour of the integrand at (or near) the point $t = 0$. Accordingly it is most usually stated in the following form:

The sampling property of the Dirac delta function: if f is any function which is continuous on a neighbourhood of the origin, then

$$\int_{-\infty}^{+\infty} f(t)\delta(t)dt = f(0). \tag{2.9}$$

In particular, let $f(t) = u(a - t)$, where a is any fixed number other than 0. Then $f(t) = 0$ for all $t > a$ and $f(t) = 1$ for all $t < a$; moreover f is certainly continuous on a neighbourhood of the origin. Hence we should have

$$\int_{-\infty}^{a} \delta(t)dt = \int_{-\infty}^{+\infty} u(a - t)\delta(t)dt = u(a), \quad \text{for any } a \neq 0.$$

This shows that if we were to assume the existence of a function $\delta(t)$ at the outset, with the sampling property (2.9), then it would at least be consistent with the role of $\delta(t)$ as a derivative, in some sense, of the unit step function: indeed, the relation $u'(t) = \delta(t)$, from which we first started, could have been deduced as a consequence of that sampling preperty. Nevertheless, it must still be remembered that the derivation given above of the sampling property is a purely formal one, and that there is as yet no evidence that any function $\delta(t)$ which exhibits such a property actually does exist.

Exercises II.

1. Show that δ behaves formally as though it denoted an even function: $\delta(-t) = \delta(t)$.

2. If a is any fixed number and f any function continuous on some neighbourhood of a, show that formally we should get

$$\int_{-\infty}^{+\infty} f(t)u'(t - a)dt = f(a).$$

3. Obtain the sampling properties which would appear to be associated with each of the following functions:

(a) the derivative of $u(-t)$;

(b) the derivative of $\text{sgn}(t)$.

2.3   THE DELTA FUNCTION AS A LIMIT

2.3.1 Failure of naive definition of $\delta(t)$. It is not difficult to show that in point of fact no function defined on the real line $\mathbb{R}$ can enjoy the properties attributed above to δ. The requirement that $\delta(t) = u'(t)$ necessarily implies that $\delta(t) = 0$ for all $t \neq 0$. Hence, for any continuous function $f(t)$, and any positive numbers ε_1, ε_2, however small, we must have

$$\int_{-\infty}^{-\varepsilon_1} f(t)\delta(t)dt = \int_{\varepsilon_2}^{+\infty} f(t)\delta(t)dt = 0$$

so that

$$\int_{-\infty}^{+\infty} f(t)\delta(t)dt = \lim_{\varepsilon_1 \downarrow 0} \int_{-\infty}^{-\varepsilon_1} f(t)\delta(t)dt$$

$$+ \lim_{\varepsilon_2 \downarrow 0} \int_{\varepsilon_2}^{+\infty} f(t)\delta(t)dt = 0.$$

This does not agree with the sampling property (2.9) except in the trivial case of a function f for which $f(0) = 0$.

Furthermore, for any $t > 0$ we actually get

$$\int_{-\infty}^{t} \delta(\tau)d\tau = \int_{-\infty}^{t} u'(\tau)d\tau = 0 \neq u(t)$$

and this is certainly not consistent with the initial assumption that $\delta(t)$ is to behave as the derivative of $u(t)$.

The derivation of the alleged sampling property given in section 2.2.2 above involved the interchange of the order of the operations of integration and of taking the limit, and the subsequent failure of $\delta(t)$ to exhibit this property shows that the interchange cannot be justified in this situation. Similarly we cannot justify the application of the usual integration by parts formula in the alternative derivation of the sampling property.

2.3.2 Simple sequential definition of $\delta(t)$. One way of dealing with this situation is to abandon the assumption that $\delta(t)$ represents an actual function, and to construct instead a suitable sequence of proper functions which will approximate the desired behaviour of $\delta(t)$ as closely as we wish. Then we could treat all expressions involving so-called delta functions in terms of a symbolic shorthand for certain limiting processes. There are many ways in which we could do this, but among the simplest and most straightforward examples are the sequences $(s_n)_{n \in \mathbb{N}}$ and $(d_n)_{n \in \mathbb{N}}$ of functions illustrated in Fig. 2.3 and defined formally below:

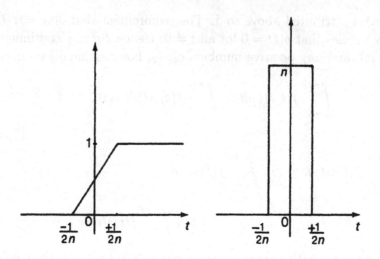

Figure 2.3:

$$s_n(t) = \begin{cases} 1 & \text{for } t > 1/2n \\ nt + 1/2 & \text{for } -1/2n \le t \le +1/2n \\ 0 & \text{for } t < -1/2n \end{cases} \qquad (2.9)$$

$$d_n(t) = \begin{cases} 0 & \text{for } t > 1/2n \\ n & \text{for } -1/2n < t < +1/2n \\ 0 & \text{for } t < -1/2n \end{cases} \qquad (2.10)$$

Then,

(i) $d_n(t) = \frac{d}{dt} s_n(t)$, for all t except $t = \pm 1/2n$,

(ii) $s_n(t) = \int_{-\infty}^{t} d_n(\tau) d\tau$, for all t,

(iii) if f is any function continuous on some neighbourhood of the origin then,

$$\int_{-\infty}^{+\infty} f(t) d_n(t) dt = n \int_{-1/2n}^{+1/2n} f(t) dt = f(\xi_n)$$

where ξ_n is some point such that $-1/2n < \xi_n < +1/2n$ (making an entirely legitimate use of the First Mean Value Theorem of the Integral Calculus).

If we allow n to tend to infinity then the sequences (s_n) and (d_n) tend in the ordinary pointwise sense to limit functions u and d_∞:

$$\lim_{n\to\infty} s_n(t) = u(t) = \begin{cases} 1 & \text{for } t > 0 \\ 0 & \text{for } t < 0 \end{cases} \qquad (2.11)$$

(In actual fact, $\lim_{n\to\infty} s_n(t) = u_{1/2}(t)$ since $s_n(0) = 1/2$ for all n.)

$$\lim_{n\to\infty} d_n(t) = d_\infty(t) = \begin{cases} +\infty & \text{for } t = 0 \\ 0 & \text{for } t \neq 0 \end{cases} \qquad (2.12)$$

and it follows that, for any function f continuous on a neighbourhood of the origin we actually get

$$\int_{-\infty}^{+\infty} f(t)d_\infty(t)dt = 0.$$

However, on the other hand we have,

$$\lim_{n\to\infty} \int_{-\infty}^{+\infty} f(t)d_n(t)dt = \lim_{n\to\infty} \int_{-1/2n}^{+1/2n} nf(t)dt$$

$$= \lim_{n\to\infty} f(\xi_n), \text{ where } -1/2n < \xi_n < +1/2n,$$

so that, by continuity,

$$\lim_{n\to\infty} \int_{-\infty}^{+\infty} f(t)d_n(t)dt = f(0).$$

Thus, for $n = 1, 2, 3, ...$, we can construct functions d_n whose operational behaviour approximates more and more closely those properties required of the so-called delta function. The symbolic expression $\delta(t)$, when it appears within an integral sign, must be understood as a convenient way of indicating that the integral in question is itself symbolic and really denotes a limit of a sequence of bona fide integrals. Further, $\delta(t)$ must be clearly distinguished from the ordinary pointwise limit function $d_\infty(t)$, which is really what equations (2.7a) and (2.7b) above specified. More precisely, when we use expressions like $\int_{-\infty}^{+\infty} f(t)\delta(t)dt$, they are to be understood simply as a convenient way of denoting the limit

$$\lim_{n\to\infty} \int_{-\infty}^{+\infty} f(t)d_n(t)dt = f(0),$$

and should not be interpreted at their face value, namely as

$$\int_{-\infty}^{+\infty} f(t)[\lim_{n\to\infty} d_n(t)]dt \equiv \int_{-\infty}^{+\infty} f(t)d_\infty(t)dt = 0.$$

2.4   STIELTJES INTEGRALS

2.4.1 Definition of Riemann-Stieltjes integral. An entirely different way of representing the sampling operation associated with the so-called delta function is to resort to a simple generalisation of the concept of integration itself. Recall that the elementary (or Riemann) theory of the integration of bounded, continuous functions over finite intervals treats the integral as the limit of finite sums

$$\int_a^b f(t)dt = \lim_{\Delta t \to 0} \sum_{k=1}^n f(\tau_k)(t_k - t_{k-1}).$$

Here the t_k are points of subdivision of the range $[a, b]$ of integration, the τ_k are arbitrarily chosen points in the corresponding sub-intervals $[t_{k-1}, t_k]$, and Δt represents the largest of the quantities $\Delta_k t \equiv t_k - t_{k-1}$. In a generalisation due to Stieltjes these elements $\Delta_k t$ are replaced by quantities of the form $\Delta_k \nu \equiv \nu(t_k) - \nu(t_{k-1})$, where ν is some fixed, monotone increasing function, and the so-called **Riemann-Stieltjes** integral appears as the limit of appropriately weighted finite sums.

Let ν be a monotone increasing function and let f be any function which is continuous on the finite, closed interval $[a, b]$. By a **partition** P of $[a, b]$ we mean a subdivision of that interval by points t_k, where $0 \le k \le n$, such that

$$a = t_0 < t_1 < ... < t_{n-1} < t_n = b.$$

For a given partition P let $\Delta_k \nu$ denote the quantity

$$\Delta_k \nu \equiv \nu(t_k) - \nu(t_{k-1}), \quad k = 1, 2, ..., n.$$

Then, if τ_k is some arbitrarily chosen point in the sub-interval $[t_{k-1}, t_k]$, where $1 \le k \le n$, we can form the sum

$$\sum_{k=1}^n f(\tau_k)\Delta_k \nu. \tag{2.10}$$

It can be shown that, as we take partitions of $[a,b]$ in which the points of subdivision, t_k, are chosen more and more closely together, so the corresponding sums (2.10) tend to some definite limiting value. This limit is called the **Riemann-Stieltjes** (RS) integral of f with respect to ν, from $t=a$ to $t=b$, and we write

$$\int_a^b f(t)d\nu(t) \equiv \lim_{\Delta t \to 0} \sum_{k=1}^n f(\tau_k)\Delta_k\nu \qquad (2.11)$$

where $\Delta t = \max(t_k - t_{k-1})$ for $1 \le k \le n$. The function $\nu(t)$ in the integral is often called the **integrator**.

In the particular case in which $\nu(t) = t$, so that $\Delta_k\nu = t_k - t_{k-1}$, the RS-integral reduces to the ordinary (Riemann) integral of f over $[a,b]$:

$$\int_a^b f(t)d\nu(t) \equiv \int_a^b f(t)dt = \lim_{\Delta t \to 0} \sum_{k=1}^n f(\tau_k)(t_k - t_{k-1}).$$

More generally, if ν is any monotone increasing function with a continuous derivative ν', then the RS-integral of f with respect to ν can always be interpreted as an ordinary Riemann integral:

$$\int_a^b f(t)d\nu(t) = \int_a^b f(t)\nu'(t)dt. \qquad (2.12)$$

2.4.2 The unit step as integrator. In contrast we consider a simple situation in which ν is discontinuous. We shall compute the value of the Stieltjes integral

$$\int_a^b f(t)du_c(t - T),$$

where f is a function continuous on the interval $[a,b]$, T is some fixed number such that $a < T < b$, and u_c is the representative of the unit step which takes the value c (where $0 \le c \le 1$) at the origin. For any given partition of $[a,b]$ there will be just two possibilities:

(i) T is an interior point of some sub-interval of $[a,b]$, so that we have $t_{k-1} < T < t_k$. In this case we will get

$$\Delta_k\nu = u_c(t_k - T) - u_c(t_{k-1} - T) = 1 - 0 = 1,$$

while

$$\Delta_r\nu = 0, \quad \text{for every } r \ne k.$$

Hence,

$$\sum_{r=1}^{n} f(\tau_r)\Delta_r \nu = f(\tau_k), \quad \text{where } t_{k-1} \leq \tau_k \leq t_k. \tag{2.13}$$

(ii) T is a boundary point of two adjacent sub-intervals, say $T = t_k$. This time we have

$$\Delta_k \nu = u_c(t_k - T) - u_c(t_{k-1} - T) = c - 0 = c,$$

$$\Delta_{k+1} \nu = u_c(t_{k+1} - T) - u_c(t_k - T) = 1 - c,$$

all other terms $\Delta_r \nu$ being zero. Hence,

$$\sum_{r=1}^{n} f(\tau_r)\Delta_r \nu = cf(\tau_k) + (1-c)f(\tau_{k+1})$$

$$= f(\tau_{k+1}) + c[f(\tau_k) - f(\tau_{k+1})] \tag{2.14}$$

where $t_{k-1} \leq \tau_k \leq T \leq \tau_{k+1} \leq t_{k+1}$.

Comparing (2.13) and (2.14), and using the fact that f is a continuous function (at least in the neighbourhood of T), it is clear that the limit to which the approximating sums tend must be $f(T)$. Note that this result is quite independent of the particular number c specified in the definition of the function u_c. (This is essentially the reason why in many contexts there is no real reason to define the unit step function u at the origin.) Further, once some specific point T has been fixed, we can always choose an interval $[a, b]$ large enough to ensure that $a < T < b$. Provided only that this condition is satisfied, we have

$$\int_a^b f(t)du_c(t - T) = f(T). \tag{2.15}$$

Since this result holds for any values of a and b such that $a < T < b$, we may take infinite limits in the integral and write the final result in a form which shows its independence of the numbers a, b, and c:

If f is any function continuous on a neighbourhood of a fixed point T, and if u denotes the unit step function (left undefined at the origin) then

$$\int_{-\infty}^{+\infty} f(t)du(t - T) = f(T). \tag{2.16}$$

In particular, if $T = 0$ **then**

$$\int_{-\infty}^{+\infty} f(t)du(t) = f(0). \tag{2.17}$$

Thus the sampling property (2.9) for arbitrary continuous functions f can be legitimately expressed in terms of a Riemann-Stieltjes integral taken with respect to the unit step function, u. Expressions involving the delta function, such as $\int_{-\infty}^{+\infty} f(t)\delta(t)dt$, may be regarded as conventional representations of this fact instead of being interpreted in terms of a limiting process as described in the preceding section. In particular the conventional use of the expression $\int_{-\infty}^{+\infty} f(t)\delta(t)dt$ as a representation of the Stieltjes integral $\int_{-\infty}^{+\infty} f(t)du(t)$ is in the spirit of equation (2.12), and gives some justification for the description of $\delta(t)$ as the 'derivative' of $u(t)$.

2.5 DEVELOPMENTS OF DELTA FUNCTION THEORY

The Stieltjes integral provides an approach to delta function theory which expresses the fundamental sampling operation, for all continuous functions, in terms of a genuine integration process. It is rather less convenient when we come to consider the meaning to be attached to derivatives of the delta function and the sampling operations which should be associated with them. A sequential definition of the delta function (and, subsequently, of its derivatives) seems to offer a much more flexible and widely applicable treatment, and as such it is very commonly adopted. In general, a **delta sequence** is a sequence of ordinary functions, $p_n(t)$, such that the characteristic sampling operation

$$f(t) \rightarrow \lim_{n \to \infty} \int_{-\infty}^{+\infty} p_n(t)f(t)dt = f(0) \tag{2.18}$$

is valid at least for a sufficiently wide class of 'test functions' $f(t)$. The sequence (d_n) used in Sec.2.3.1 is a particularly simple example of such a delta sequence, as also are the sequences (g_n) and (h_n) defined in Exercises III below. More generally, if $p(t)$ is some fixed, non-negative function such that

$$\int_{-\infty}^{+\infty} p(t)dt = 1 \tag{2.19}$$

and for $n = 1, 2, 3, \ldots$ we define

$$p_n(t) = np(nt) \tag{2.20}$$

then it can be shown that (p_n) is a delta sequence giving the required sampling operation for all bounded continuous functions $f(t)$.

However, powerful and widely applicable though the delta sequence method may be, the systematic use of infinite sequences and limiting operations is likely to seem rather ponderous to the student encountering delta function calculus for the first time. Accordingly we adopt, in the first part of this book, a more naive, direct approach, only rarely appealing to either the sequential definition or to that in terms of a Stieltjes integral. As explained more fully in the next chapter, we shall concentrate instead on developing a consistent formal calculus which tells us how the symbol δ is to be used when it occurs in various algebraic contexts. Once this has been established and the usage has become familiar it will be appropriate to consider more fully the possible interpretation of the formal symbolism.

Exercises III.

1. By direct evaluation of the integral $\int_{-\infty}^{+\infty} \cos(t)d_n(t)dt$, where the functions $d_n(t)$ are as defined in Sec. 2.3, confirm that

$$\int_{-\infty}^{+\infty} \cos(t)\delta(t)dt = 1,$$

in the symbolic sense attributed to integrals involving the delta function.

2. Let (g_n) be a sequence of functions defined as follows:

$$g_n(t) = \begin{cases} n^2t + n & \text{for } -1/n \leq t < 0 \\ n - n^2t & \text{for } 0 \leq t < 1/n \\ 0 & \text{for all other values of } t \end{cases} \tag{2.13}$$

Show that, in the pointwise sense, $\lim_{n\to\infty} g_n(t) = d_\infty(t)$, and repeat the computation of question 1, with the functions g_n in place of the functions d_n. Sketch the graph of a typical function G_n, defined by

$$G_n(t) = \int_{-\infty}^{t} g_n(\tau)d\tau.$$

3. Let (h_n) be a sequence of functions defined as follows:

$$h_n(t) = \begin{cases} 2n^2t & \text{for } 0 \le t < 1/2n \\ 2n - 2n^2t & \text{for } 1/2n \le t < 1/n \\ 0 & \text{for } 1/n \le t \end{cases} \qquad (2.14)$$

and, $h_n(-t) = h_n(t)$.

Show that it is not true that $\lim_{n \to \infty} h_n(t) = d_\infty(t)$ in the pointwise sense. Repeat the computation of question 1 with the functions h_n in place of the functions d_n, and sketch the graph of a typical function, H_n, defined by

$$H_n(t) = \int_{-\infty}^{t} h_n(\tau) d\tau.$$

4. If f is continuous on the closed interval $[a, b]$ and if c is any number such that $0 \le c \le 1$, show from first principles that

$$\text{(a)} \quad \int_{a}^{b} f(t) du_c(t - a) = (1 - c)f(a)$$

and

$$\text{(b)} \quad \int_{a}^{b} f(t) du_c(t - b) = cf(b).$$

2.6 HISTORICAL NOTE

Although the unit step function is usually associated with the name of Heaviside and the delta function with that of Dirac, both concepts can be found earlier in the literature. Cauchy uses the unit step under the name 'coefficient limitateur' and defines it by the formula

$$u(t) = \frac{1}{2} \left[1 + \frac{t}{\sqrt{t^2}} \right].$$

Moreover Cauchy in 1816 (and, independently, Poisson in 1815) gave a derivation of the Fourier integral theorem by means of an argument involving what we would now recognise as a sampling operation of the type associated with a delta function. And there are similar examples of the use of what are essentially delta functions by Kirchoff, Helmholtz, and, of course, Heaviside himself. But it is undeniable that Dirac was the first to use the notation $\delta(t)$ and to state unequivocally the more important properties which should be associated with the delta function. In **The Principles of Quantum Mechanics** (1930) Dirac refers to δ as an 'improper function', and makes it quite clear that he is defining a mathematical entity of a new type, and one which cannot be identified with an ordinary function. For Dirac it is the application to continuous functions of what we have called the sampling property which is the central feature of his treatment of the delta function, and he derives this by sketching a limiting process such as that used in our Sec.2.2.1. The alternative derivation using a formal integration by parts, is appropriate only in respect of a sampling property which applies to smooth functions. In many applications this distinction is of minor importance, but it can be of some real significance in the context of Fourier analysis, as will be seen in the sequel. From now on we will use the term **Dirac delta function** whenever the sampling operation in question applies to all continuous functions, as Dirac originally specified.

It is fair to say that no account of the basic theory of the delta function which is both reasonably comprehensive and mathematically satisfactory was generally available until the publication in 1953 of the **Theory of Distributions** by Laurent Schwartz. (The work of Sobolev which in some respects genuinely antedates that of Schwartz did not become widely known until later.) To do full justice to the ideas of Schwartz would demand a fairly extensive acquaintance on the part of the reader with the concepts and processes of modern analysis, and for the moment we shall take the subject no further. In the chapters which follow we adopt a philosophy which might well be attributed to Dirac himself. That is to say we recognise (1) the central importance of the so-called sampling property, and (2) the fact that the use of the symbols $\delta(t)$, $\delta(t-a)$ and so on, to represent that property is, at best, a purely formal convenience. Our intention in Chapters 3 to 6 is to devise systematic rules of procedure governing the use of such symbols in situations in which the normal operations of algebra and of elementary calculus may be called into play. The reader should then be sufficiently familiar with the properties and applications of delta functions to benefit from a more rigorous approach.

An introduction to other types of commonly occurring generalised functions will be given in Chapter 7, followed by a sketch of the comprehensive theory of (linear) generalised functions developed by Laurent Schwartz. It is hoped that this will enable the reader to consult one or other of the many standard texts now available on the Schwartz theory of distributions. In the final chapters of this book we offer an alternative approach to the whole subject area, using some of the ideas and methods of what is now called **Nonstandard Analysis (NSA)**. This allows a treatment of the delta function, and of other generalised functions, which is both intuitively appealing and entirely rigorous. Moreover, since such an interpretation of generalised functions is one of the more immediately effective illustrations of the power of nonstandard methods, it is hoped that the reader may be encouraged to explore the use of NSA in other contexts.

Further reading.

The definition by Dirac himself of his "improper function", $\delta(t)$, together with a description of some of its more essential properties, is given succinctly and quite clearly in Chapter III of his **Principles of Quantum Mechanics** (Third ed., Oxford, 1947). Arguably what he says there ought to be sufficient for any reader with adequate knowledge of classical analysis to understand the significance of the notation which Dirac introduces, and to make successful and trouble-free use of it.

But, as many students have found to their cost, the conceptual basis on which the delta function is supposed to be founded is far from clear, and a more rigorous basic approach is essential to prevent misunderstanding. The most popular, and perhaps the most easily grasped, theory of the delta function is to treat it as, in a certain sense, the limit of a sequence of ordinary functions. What is more, this kind of approach can be readily extended to allow the definition of a wide range of other generalised functions, over and above the delta function itself, and its derivatives. It is given a straightforward and comprehensive treatment in the justly celebrated book, **An Introduction to Fourier Analysis and Generalised Functions** (C.U.P. 1960) by **Sir James Lighthill**. This is strongly recommended as supplementary reading, although it may be found somewhat more demanding than the treatment given here.

A mine of general information on the theory, applications and history of the delta function is to be found in the classic text **Operational**

Calculus based on the two-sided Laplace integral by **Balth. van der Pol** and **H.Bremmer** (C.U.P. 1955). Although published over 50 years ago this is still a most valuable reference text and is again strongly recommended for supplementary reading. However, the reader should perhaps be warned that this book does, unfortunately, introduce yet another variant notation for the Heaviside unit step function; this appears as $U(t)$ and, more importantly, is constrained to assume the value $\frac{1}{2}$ at the origin. In addition to many interesting and useful examples of delta sequences, it does also contain some discussion of the less well known representation of the delta function as a Stieltjes integral.

A little later there appeared in **Modern Mathematics for the Engineer** (McGraw-Hill, 1961), edited by **Edwin F. Beckenbach**, a famous article by **Arthur Erdélyi** called "From Delta Functions to Distributions". Apart from a brief historical survey of the delta function itself, this gives a succinct, but very clear, account of an important alternative approach to the whole topic of generalised functions which space does not allow us to discuss in the present book. This is the theory of operational calculus developed by **Jan Mikusinski**.

A careful and thorough account of the elementary Stieltjes integral, used as the basis of a restricted but entirely rigorous theory of the delta function is to be found in **Mathematics of Dynamical Systems** (Nelson, 1970) by **H.H.Rosenbrock** and **C.Storey**.

For those interested in more detailed historical aspects of the emergence of the concept of generalised functions, and of the delta function in particular, there is the very readable account contained in **George Temple's** delightful book **100 Years of Mathematics** (Duckworth, 1981). But even this must give pride of place to the comprehensively researched study by **Jesper Lüzen** in **The Pre-history of the Theory of Distributions** (Springer-Verlag, 1982).

Chapter 3

Properties of the Delta Function

3.1 THE DELTA FUNCTION AS A FUNCTIONAL

3.1.1 Functions and functionals. Recall that, as stated in Sec. 1.2.1, we reserve the term "function" to denote a real, single-valued function of a single real variable. That is to say, we understand a function to be a mapping of (real) numbers into numbers, and therefore to be completely characterised as such. It has been made clear in Chapter 2 that the so-called delta function is not a function in this strict sense at all, and is not determined by the values which may be attributed to $\delta(t)$ as t ranges from $-\infty$ to $+\infty$. The situation can be made somewhat clearer if we consider first the sense in which we are really to understand the Heaviside unit step:

For each given value of the parameter c the symbol u_c denotes a proper function which has a well-defined pointwise specification over the entire range $-\infty < t < +\infty$. Now let f be an arbitrary bounded continuous function which vanishes identically outside some finite interval. Then we can write

$$\int_{-\infty}^{+\infty} f(t)u_c(t)dt = \int_{0}^{+\infty} f(t)dt. \qquad (3.1)$$

(The constraints on f in (3.1) are simply to ensure the existence of the

47

integrals in (3.1) and could obviously be weakened considerably.) Equation (3.1) expresses a certain operational property associated with u_c in the sense that it defines a mapping or transformation: to each continuous function f which vanishes outside some finite interval there corresponds a well-defined number $\int_0^{+\infty} f(t)dt$. This number is independent of the particular value $c = u_c(0)$ and hence of the particular function $u_c(t)$. Recalling our convention that u is to stand for the unit step function left undefined at the origin, we could equally well write (3.1) in the form,

$$\int_{-\infty}^{+\infty} f(t)u(t)dt = \int_0^{+\infty} f(t)dt. \tag{3.2}$$

The symbol u appearing on the left-hand side of (3.2) may now be given an alternative interpretation. Instead of standing for some one particular function, it could be regarded as the representative of an entire family of equivalent functions, u_c, any one of which would suffice to characterise the specific operation on f which we have in mind. From this point of view we have no real need to specify the precise pointwise behaviour of the unit step. It is enough that the symbol u appearing in (3.2) is known to define a certain mapping of a specific class of functions into numbers, and we could indicate this by adopting an appropriate notation:

To each continuous function f which vanishes outside some finite interval there corresponds a certain number which we usually write as $u(f)$, or sometimes as $< u, f >$, and which is given by the mapping

$$f \to u(f) \equiv < u, f > = \int_0^{+\infty} f(t)dt. \tag{3.3}$$

In the same way the one important feature of the so-called delta function, $\delta(t)$, is that it provides a convenient means of representing an operation, defined at least for all functions continuous on a neighbourhood of the origin, which maps or transforms each such function f into the value $f(0)$ which it assumes at the origin. Following the convention suggested in (3.3) we could adopt a corresponding notation, $\delta(f)$ or $< \delta, f >$, and write this operation in the form

$$f \to \delta(f) \equiv < \delta, f > = f(0).$$

Symbols such as u and δ when regarded as specifying operations on certain classes of functions (rather than as supposedly standing for actual functions $u(t)$ and $\delta(t)$ in their own right) are properly referred to as **functionals** - a term which we will treat more fully in Chapter 8 below.

However, it is usually more agreeable, and often more convenient, to retain the familiar notation of the integral calculus and to write the mappings defined by u and δ in the form

$$f \rightarrow \int_0^{+\infty} f(t)dt = \int_{-\infty}^{+\infty} f(t)u(t)dt. \tag{3.4a}$$

and

$$f \rightarrow f(0) = \int_{-\infty}^{+\infty} f(t)\delta(t)dt \tag{3.4b}$$

The significance of the integral $\int_{-\infty}^{+\infty} f(t)u(t)dt$ in (3.4a) then appears primarily as its capacity to express the number $u(f)$ in a familiar canonical form, although a genuine integration process is indeed involved. In (3.4b) however, both the integral sign and the expression $\delta(t)$, which seems to imply existence of an actual function, must be understood in a purely symbolic sense. This use of conventional notation, appropriately borrowed from elementary calculus, is perhaps the main reason why u, δ, and, as we shall see later, certain other functionals, are more often described as **generalised functions** - possibly a somewhat misleading term but one which is now very generally accepted.

In what follows we shall develop a set of rules which will allow us to interpret expressions involving u and δ (and other, allied, symbols denoting generalised functions) and to apply to them the usual processes of elementary calculus. Throughout it is only the operational significance of u and δ which will concern us. All the same, the term "function" is quite appropriate to the unit step, u, since its operational behaviour is adequately specified by the pointwise definition of $u(t)$. In the case of δ this is not so and strictly only the notation $< \delta, f >$, or $\delta(f)$, makes sense. Once again it must be emphasised that no inferences can safely be drawn from the apparent pointwise behaviour attributed to $\delta(t)$ although, on occasion, such inferences may turn out to be correct. Nevertheless it is often helpful to think of $\delta(t)$ as though it denoted a function in the ordinary sense which takes the value zero everywhere except at the origin where its value may be presumed infinite; i.e. to think that $\delta(t) \equiv$

$d_\infty(t)$. Provided that this pointwise specification is treated as purely descriptive and is not used to justify algebraic manipulations etc., the abuse of notation involved is harmless (and, indeed, too well established to be wholly ignored). We shall therefore continue to write $\delta(t)$ and to refer to it as "the delta function" even though we should more properly speak of "the delta functional".

3.1.2 Translates. For an ordinary function g and a fixed real number a the symbol g_a is often used to denote the **translate** of g with respect to a:

$$g_a(t) \equiv g(t-a), \qquad (3.5)$$

and for any other function f we can write

$$\int_{-\infty}^{+\infty} f(t)g_a(t)dt \equiv \int_{-\infty}^{+\infty} f(t)g(t-a)dt = \int_{-\infty}^{+\infty} f(\tau + a)g(\tau)d\tau$$

so that formally we should have

$$\int_{-\infty}^{+\infty} f(t)\delta(t-a)dt = \int_{-\infty}^{+\infty} f(\tau + a)\delta(\tau)d\tau$$

$$= [f(\tau + a)]_{\tau=0} = f(a).$$

This symbolic calculation suggests that the translate of the delta function, δ_a, should represent, or characterise, the operation which carries any function f, continuous on a neighbourhood of a, into the value $f(a)$ which that function assumes at a. Moreover the evaluation of a Stieltjes integral of the form

$$\int_{-\infty}^{+\infty} f(t)du(t-a)$$

(as in Sec. 2.4) is enough to show that δ_a may be regarded as the formal derivative of the translated step function $u(t-a)$. Hence, as a more general starting point, we could take the generalised function δ_a as that which is defined by the sampling operation

$$f \to \delta_a(f) \equiv f(a),$$

where f is any function which is continuous on some neighbourhood of a. Symbolically we write

$$\int_{-\infty}^{+\infty} f(t)\delta_a(t)dt \equiv \int_{-\infty}^{+\infty} f(t)\delta(t-a)dt = f(a)$$

and,

$$\frac{d}{dt}u(t-a) = \delta(t-a). \tag{3.6}$$

3.1.3 Remark. In the case of the unit step function it is not convenient to make use of the conventional notation for the translate. This is because of the special significance of the subscript c in $u_c(t)$, as described in Sec.2.1.2. Therefore we shall always denote a translate of the unit step function by writing it in full, as $u_c(t-a)$.

3.2 SUMS AND PRODUCTS

3.2.1 Sums involving delta functions. We require to assign meaning to combinations like $h + \delta$, where h is an ordinary function, and to expressions of the form $\delta_a + \delta_b$. If h_1 and h_2 are any two ordinary functions then their **pointwise sum** is defined as the function h_3 whose value at each point t is the sum of the values assumed by h_1 and h_2 at that point:

$$h_3(t) \equiv (h_1 + h_2)(t) = h_1(t) + h_2(t). \tag{3.7}$$

For delta functions we cannot appeal to pointwise behaviour so that (3.7) is of no immediate relevance as it stands. However, for any other function f for which all the integrals involved exist we certainly have

$$\int_{-\infty}^{+\infty} f(t)h_3(t)dt = \int_{-\infty}^{+\infty} f(t)[h_1(t) + h_2(t)]dt$$

$$= \int_{-\infty}^{+\infty} f(t)h_1(t)dt + \int_{-\infty}^{+\infty} f(t)h_2(t)dt, \tag{3.8}$$

and we can generalise (3.8) in an obvious way. For any function f, continuous on a neighbourhood of the origin, we may write by analogy with (3.8),

$$\int_{-\infty}^{+\infty} f(t)[h(t) + \delta(t)]dt = \int_{-\infty}^{+\infty} f(t)h(t)dt + \int_{-\infty}^{+\infty} f(t)\delta(t)dt$$

$$= \int_{\infty}^{+\infty} f(t)h(t)dt + f(0). \tag{3.9}$$

Similarly, provided that f is continuous on a neighbourhood of a and on a neighbourhood of b,

$$\int_{-\infty}^{+\infty} f(t)[\delta_a + \delta_b](t)dt = \int_{-\infty}^{+\infty} f(t)\delta_a(t)dt + \int_{-\infty}^{+\infty} f(t)\delta_b(t)dt$$

$$= f(a) + f(b). \qquad (3.10)$$

Equations (3.9) and (3.10) effectively define sampling properties which characterise $h+\delta$ and $\delta_a+\delta_b$ respectively as generalised functions. These results clearly depend on the assumption that the distributive law of ordinary algebra should remain valid in expressions involving delta functions and ordinary (continuous) functions.

3.2.2 Products with continuous functions. Now consider the formal product $\phi\delta$, where ϕ is a function continuous at least on some neighbourhood of the origin. If f is any other such function, and we assume that the associative law of multiplication continues to hold for the factors of the integrands in the symbolic integrals concerned, then we should have

$$\int_{-\infty}^{+\infty} f(t)[\phi(t)\delta(t)]dt = \int_{-\infty}^{+\infty} [f(t)\phi(t)]\delta(t)dt = f(0)\phi(0). \qquad (3.11)$$

In particular if ϕ is a constant function, say $\phi(t) = k$ for all t, then we should get

$$\int_{-\infty}^{+\infty} f(t)[k\delta(t)]dt = \int_{-\infty}^{+\infty} [kf(t)]\delta(t)dt$$

$$= k\int_{-\infty}^{+\infty} f(t)\delta(t)dt = kf(0). \qquad (3.12)$$

Note that we may write the result (3.11) in the equivalent form

$$\phi(t)\delta(t) = \phi(0)\delta(t) \qquad (3.13)$$

in the sense that the the formal product $\phi(t)\delta(t)$ has the same operational significance as that given to $k\delta(t)$ by equation (3.12) when the constant k has the value $\phi(0)$. It is customary to refer to $k\delta(t)$ as a delta function of **strength** k. For $k = 0$ we have, in particular,

$$\int_{-\infty}^{+\infty} f(t)[0\delta(t)]dt = 0f(0) = 0$$

for all (continuous) functions f, so that the product $0\delta(t)$ has the same operational significance as the function which vanishes identically on $\mathbb{R}$. Accordingly it makes sense to write $0\delta(t) = 0$: a delta function of strength zero is null.

Caution: It is perhaps worthwhile at this stage to see what a naive argument based on the pointwise description of $\delta(t)$ might have led us to believe. If we were to identify $\delta(t)$ with the ordinary function $d_\infty(t)$ then for any positive number k we would have

$$k\delta(t) \equiv kd_\infty(t) = \begin{cases} +\infty & \text{for } t = 0 \\ 0 & \text{for } t \neq 0 \end{cases}$$

This would suggest that $k\delta(t) = \delta(t)$, a result incompatible with the operational property of the product $k\delta(t)$ expressed by equation (3.10), except in the particular case when $k = 1$. Once again it is clear that δ represents something other than the ordinary function $d_\infty(t)$ which vanishes for all $t \neq 0$ and takes the value $+\infty$ at the origin.

3.2.3 Discontinuous multiplier. In considering products involving delta functions we have been careful to confine attention to continuous multipliers. This is because the characteristic sampling property of the Dirac delta function is itself only defined for continuous functions (or, at most, for those functions which are continuous on some neighbourhood of the origin). If $f(t)$ has only a removable discontinuity at the origin then a trivial extension of the sampling property is all that is required, and we may write

$$\int_{-\infty}^{+\infty} f(t)\delta(t)dt = \lim_{t \to 0} f(t).$$

It is a different matter if there is a jump discontinuity at the origin. It would, for example, be convenient for a number of applications to assign a meaning to the product $u\delta$ of the unit step and the delta function. One way in which this might be done can be described as follows. Suppose that f_0 is some fixed, continuous, function such that $f_0(0) = 1$. For an arbitrary continuous function f let $\hat{f}$ denote the function defined by

$$\hat{f}(t) = f(t) - f(0)f_0(t).$$

Since $\hat{f}$ is clearly continuous and such that $\hat{f}(0) = 0$, it follows that the product $u_c(t)\hat{f}(t)$ is defined for all t and is continuous everywhere and independent of the number c. Hence we may write

$$\int_{-\infty}^{+\infty} \hat{f}(t)[u(t)\delta(t)]dt \equiv \int_{-\infty}^{+\infty} [\hat{f}(t)u_c(t)]\delta(t)dt = \hat{f}(0)c = 0.$$

Once again assuming that all the products involved obey the usual associative and distributive laws this gives

$$0 = \int_{-\infty}^{+\infty} [f(t) - f(0)f_0(t)][u(t)\delta(t)]dt$$

$$= \int_{-\infty}^{+\infty} f(t)[u(t)\delta(t)]dt - f(0)\int_{-\infty}^{+\infty} f_0(t)[u(t)\delta(t)]dt,$$

that is

$$\int_{-\infty}^{+\infty} f(t)[u(t)\delta(t)]dt = f(0)\int_{-\infty}^{+\infty} f_0(t)[u(t)\delta(t)]dt. \tag{3.14}$$

Thus we could define the sampling operation characteristic of the formal product $u\delta$ for all continuous functions f provided only that it is well defined in respect of one particular such function f_0. In fact from (3.14) we would conclude that

$$u(t)\delta(t) = k\delta(t) \tag{3.15}$$

where k is an arbitrary constant.

This is as far as we can go with any certainty. Admittedly there are some plausible arguments which suggest that the most appropriate value for k in (3.15) is 1/2. For example a formal differentiation of the function $u^2(t)$ gives

$$\frac{d}{dt}u^2(t) = 2u(t)\frac{d}{dt}u(t) = 2u(t)\delta(t) = 2k\delta(t).$$

But, clearly, we have $u^2(t) = u(t)$ so that

$$\frac{d}{dt}u^2(t) = \frac{d}{dt}u(t) = \delta(t),$$

which implies that $u(t)\delta(t) = \frac{1}{2}\delta(t)$. However, other considerations suggest that the choice of $k = 1/2$ in (3.15) may not be appropriate in every context. For example, making what would seem to be the very reasonable assumption that the formal product of generalised functions should be associative, we have on the one hand

$$[\delta(t)u(t)]u(t) = k\delta(t)u(t) = k^2\delta(t)$$

while on the other, since $u^2(t) = u(t)$,

$$\delta(t)[u(t)u(t)] = \delta(t)u(t) = k\delta(t).$$

This implies that $k^2 = k$ so that it appears that $k = 0$ or $k = 1$ are possible choices for the constant in (3.15), but not $k = 1/2$.

It is clear that the definition of multiplication for generalised functions is by no means a straightforward matter, and that it can present very real problems. Where ordinary functions are concerned we expect to find that products satisfy the usual fundamental laws of algebra; that is to say, they are commutative, associative and distributive. It appears that this may not be so when we try to handle products of generalised functions. The significance of the product $u(t)\delta(t)$ in particular remains an open question, and we shall be obliged to consider it more fully in Sec.3.6 below, and again in Chapter 5 in connection with the problem of defining the Laplace Transform of the delta function.

Exercises I.

1. Find formal simplified equivalent expressions for:

(a) $\cos(t) + \sin(t)\delta(t)$; (b) $\sin(t) + \cos(t)\delta(t)$; (c) $1 + 2e^t\delta(t - 1)$.

2. Evaluate the following (symbolic) integrals:

(a) $\int_{-\infty}^{+\infty}(t^2 + 3t + 5)\delta(t)dt$;   (b) $\int_{-\infty}^{+\infty}\frac{\cos(x)\delta(x)}{2e^x+1}dx$;

(c) $\int_{-\infty}^{+\infty}e^{-t^2}\delta(t - 2)dt$;   (d) $\int_{-\infty}^{+\infty}\sinh(2t)\delta(2 - t)dt$;

(e) $\int_{-\infty}^{+\infty}\frac{1+e^{2x}\delta(x+1)}{1+x^2}dx$;   (f) $\int_{-\infty}^{+\infty}t^2\sum_{k=1}^{n}\delta(t - k)dt$;

(g) $\int_{-\infty}^{+\infty}[e^{2\theta}u(-\theta) - \cos(2\theta)\delta(\theta) + e^{-2\theta}u(\theta)]d\theta$;

(h) $\int_{-\infty}^{+\infty}[e^{\tau}\delta_{\pi}(\tau) + e^{(-\tau)}\delta_{-\pi}(\tau)]d\tau$.

3.3 DIFFERENTIATION

3.3.1 Differentiation at jump discontinuities. We can now attach meaning to the term 'derivative' in the case of a function which has one or more jump discontinuities. Let ϕ_1 and ϕ_2 be continuously differentiable functions and let f be defined as the function which is equal to ϕ_1 for all $t < a$ and equal to ϕ_2 for all $t > a$:

$$f(t) = \phi_1(t)u(a - t) + \phi_2(t)u(t - a).$$

The function $f(t)$ may not be defined at the point $t = a$ itself, but we surely have $f(a-) = \phi_1(a)$ and $f(a+) = \phi_2(a)$.

Then,

$$\frac{d}{dt}f(t) = u(a - t)\frac{d}{dt}\phi_1(t) + \phi_1(t)\frac{d}{dt}u(a - t)$$

$$+ u(t - a)\frac{d}{dt}\phi_2(t) + \phi_2(t)\frac{d}{dt}u(t - a).$$

Now,

$$\frac{d}{dt}u(t - a) = \frac{du(t - a)}{d(t - a)}\frac{d(t - a)}{dt} = \delta(t - a) \equiv \delta_a(t)$$

and

$$\frac{d}{dt}u(a - t) = \frac{du(a - t)}{d(a - t)}\frac{d(a - t)}{dt} = -\delta(a - t) \equiv -\delta_a(t).$$

Hence,

$$\frac{d}{dt}f(t) = \phi_1'(t)u(a - t) + \phi_2'(t)u(t - a) - \phi_1(t)\delta_a(t) + \phi_2(t)\delta_a(t)$$

$$= \phi_1'(t)u(a - t) + \phi_2'(t)u(t - a) + [\phi_2(a) - \phi_1(a)]\delta_a(t). \qquad (3.16)$$

The discontinuity at $t = a$ thus gives rise to a delta function at $t = a$ multiplied by the saltus of the function f at that point, that is to say, by the number $k = f(a+) - f(a-)$. In other words we have a delta function of strength k at $t = a$. This means that integration of (3.16) will reintroduce the jump discontinuity in the primitive, f.

In particular, suppose that $\phi_1(t) = \phi(t)$ and $\phi_2(t) = \phi(t) + k$. Then (3.16) becomes

$$\frac{d}{dt}f(t) = \phi'(t)u(a - t) + \phi'(t)u(t - a) + [(\phi(a) + k) - \phi(a)]\delta_a(t)$$

$$= \phi'(t) + k\delta_a(t).$$

Further, if x and t are any two numbers such that $x < a < t$ then we get

$$\int_x^t \frac{d}{d\tau}f(\tau)d\tau = \int_x^a \phi'(\tau)d\tau + \int_a^t \phi'(\tau)d\tau + k\int_x^t \delta(\tau - a)d\tau$$

$$= \phi(a) - \phi(x) + \phi(t) - \phi(a) + k = f(t) - f(x),$$

the integral involving the delta function at $t = a$ being well-defined precisely because we have $x < a < t$. (See Sec.3.6.2 below.)

3.3.2 Classical and generalised derivatives. This extension of the concept of differentiation does give rise to certain notational difficulties. It is standard practice to denote by f' the derivative of a function f even when that derivative is not defined everywhere. For example, let $f(t) = |t|$, and write $\phi(t) = \text{sgn}(t)$. Then $f(t)$ is differentiable everywhere except at the origin, and we always have $f'(t) = \phi(t)$, for $t \neq 0$. Further, no matter what value we assign to $\phi(0)$, it is always the case that

$$\int_{-1}^t \phi(\tau)d\tau = \int_{-1}^t f'(\tau)d\tau = |t| - 1 = f(t) - f(-1).$$

Now consider the function ϕ itself. This is again a function differentiable (in the usual sense) everywhere except at the origin. If we adopt the usual conventions of elementary calculus then we would write $\phi'(t) = 0$, for all $t \neq 0$. Moreover, with the ordinary interpretation of the integral sign, it would be entirely correct to write

$$\int_{-1}^t \phi'(\tau)d\tau = \int_{-1}^t 0 d\tau = 0.$$

However, we now wish to be able to identify $2\delta(t)$ in this case as, in some sense, the 'derivative' of $\phi(t)$. This would ensure that a formal integration will recover the original discontinuous function, $\text{sgn}(t)$, at least up to an arbitrary constant. Some mode of distinguishing between the elementary concept of derivative and this new, generalised sense (which may include delta functions) seems called for. Usually, of course, the context should make clear which is the intended meaning. When there is danger of real confusion on this point we can always use the following convention:

if

$$f(t) = \phi_1(t)u(a - t) + \phi_2(t)u(t - a)$$

as above, then

$$f'(t) = \phi_1'(t)u(a - t) + \phi_2'(t)u(t - a), \quad \text{for } t \neq a,$$

but

$$Df(t) = \phi_1'(t)u(a - t) + \phi_2'(t)u(t - a) + [\phi_2(a) - \phi_1(a)]\delta_a(t),$$

that is

$$Df(t) \equiv f'(t) + [f(a+) - f(a-)]\delta(t - a). \tag{3.17}$$

We shall call $f'(t)$ the **classical** derivative of $f(t)$, and refer to $Df(t)$ as the **generalised** or the **operational** derivative of $f(t)$. Thus, for example, the classical derivative, $u'(t)$, of the unit step function $u(t)$ is the function which is equal to 0 for all $t \neq 0$ and is undefined at the origin (or else is assigned the conventional value $+\infty$ there):

$$u'(t) = d_\infty(t).$$

The generalised derivative, $Du(t)$, of the unit step is the delta function, $\delta(t)$:

$$Du(t) = \delta(t).$$

Exercises II.

1. Find the first (generalised) derivative of:

(a) $u(1 + t) + u(1 - t)$; (b) $[1 - u(t)]\cos(t)$;

(c) $[u(t - \pi/2) - u(t - 3\pi/2)]\sin(t)$;

(d) $[u(t + 2) + u(t) - u(t - 2)]e^{-2t}$.

2. Find the first and second (generalised) derivatives of:

(a) $|t|$; (b) $e^{-|t|}$; (c) $\sin(|t|)$.

3*. Show that if $f(t) = \tanh(1/t)$, for $t \neq 0$, then f has a jump discontinuity at the origin and find the (generalised) first derivative of f. Show
also that if we write

$$g(t) = \frac{d}{dt} \tanh\left(\frac{1}{t}\right), \quad \text{for } t \neq 0, \text{ and } g(0) = 0,$$

then g is a function which is continuous for all t.

3.4 DERIVATIVES OF THE DELTA FUNCTION

3.4.1 Definition of $\delta'(t)$. To obtain a meaningful 'derivative' for the
delta function we have to define a generalised function with an appropriate sampling property. For any number $a \neq 0$ the generalised function
$\frac{1}{a}[\delta(t+a) - \delta(t)]$ certainly has a well-defined sampling property, and we
need to examine what happens to this in the limit as a tends to zero.
Suppose, then, that f is any function which is continuously differentiable
in some neighbourhood of the origin. If a is sufficiently small in absolute
magnitude, we have:

$$\int_{-\infty}^{+\infty} f(t) \left[\frac{\delta(t+a) - \delta(t)}{a} \right] dt$$

$$= \frac{1}{a} \left[\int_{-\infty}^{+\infty} f(t)\delta(t+a)dt - \int_{-\infty}^{+\infty} f(t)\delta(t)dt \right]$$

$$= \frac{1}{a}[f(-a) - f(0)].$$

Hence

$$\lim_{a \to 0} \int_{-\infty}^{+\infty} f(t) \left[\frac{\delta(t+a) - \delta(t)}{a} \right] dt = -f'(0).$$

This suggests that $\delta'(t)$ should be characterised by the following property:

**If f is any function which has a continuous derivative f', at least
in some neighbourhood of the origin, then**

$$\int_{-\infty}^{+\infty} f(\tau)\delta'(\tau)d\tau = -f'(0). \tag{3.18}$$

This might have been inferred from a formal integration by parts:

$$\int_{-\infty}^{+\infty} f(t)\delta'(t)dt = [f(t)\delta(t)]_{-\infty}^{+\infty} - \int_{-\infty}^{+\infty} \delta(t)f'(t)dt$$

$$= [f(0)\delta(t)]_{-\infty}^{+\infty} - f'(0) = -f'(0).$$

Strictly this does involve a tacit appeal to the pointwise behaviour of the delta function. However, all that we are really using is the fact that the operational property of δ is confined absolutely to the point $t = 0$. At all other points it has no effect, and this is what is really meant by saying that it 'evaluates to zero' at such points.

A straightforward generalisation of the argument leading to (3.18) gives the following rule for the sampling property associated with the nth derivative of the delta function:

For each given positive integer n the generalised function $D^n\delta \equiv \delta^{(n)}$, (the nth derivative of the delta function), is defined by the characteristic property

$$\int_{-\infty}^{+\infty} f(\tau)\delta^{(n)}(\tau)d\tau = (-1)^n f^{(n)}(0) \tag{3.19}$$

where f is any function with continuous derivatives at least up to the nth order in some neighbourhood of the origin.

3.4.2 Properties of $\delta'(t)$. For the most part the properties of the derivatives $\delta^{(n)}$ are fairly obvious generalisations of the corresponding properties of the delta function itself, and can be established by using similar arguments. We list them briefly here, enlarging on those of particular difficulty and/or importance.

(i) **Translation.** For any continuously differentiable function f and any scalar a we have

$$\int_{-\infty}^{+\infty} f(\tau)\delta_a'(\tau)d\tau \equiv \int_{-\infty}^{+\infty} f(\tau)\delta'(\tau - a)d\tau$$

$$= \int_{-\infty}^{+\infty} f(t+a)\delta'(t)dt = [-f'(t+a)]_{t=0} = -f'(a). \tag{3.20}$$

More generally,

$$\int_{-\infty}^{+\infty} f(\tau)\delta_a^{(n)}(\tau)d\tau \equiv \int_{-\infty}^{+\infty} f(\tau)\delta_a^{(n)}(\tau - a)d\tau = (-1)^n f^{(n)}(a) \tag{3.21}$$

for any function f which has continuous derivatives at least up to the nth order in some neighbourhood of the point $t = a$.

(ii) **Addition.** Let n and m be positive integers. If f is any function which has continuous derivatives at least up to the nth order on a neighbourhood of the point $t = a$, and at least up to the mth order in some neighbourhood of the point $t = b$, then

$$\int_{-\infty}^{+\infty} f(\tau)[\delta_a^{(n)}(\tau) + \delta_b^{(m)}(\tau)]d\tau$$

$$= (-1)^n f^{(n)}(a) + (-1)^m f^{(m)}(b). \tag{3.22}$$

(iii) **Multiplication by a scalar.** The product $k\delta^{(n)}$, where k is any fixed number, has a well defined sampling operation given by the following equivalence:

$$\int_{-\infty}^{+\infty} f(\tau)[k\delta^{(n)}(\tau)]d\tau = \int_{-\infty}^{+\infty} [kf(\tau)]\delta^{(n)}(\tau)d\tau$$

This has the value $(-1)^n k f^{(n)}(0)$. In particular if $k = 0$ then we may write $k\delta^{(n)}(t) = 0$.

(iv) **Multiplication by a function.** If ϕ is a fixed function which is continuously differentiable in a neighbourhood of the origin, then a meaning can be given to the product $\phi(t)\delta'(t)$. If we assume that associativity holds for all the products involved then we ought to have

$$\int_{-\infty}^{+\infty} f(\tau)[\phi(\tau)\delta'(\tau)]d\tau = \int_{-\infty}^{+\infty} [f(\tau)\phi(\tau)]\delta'(\tau)d\tau$$

$$= -\left[\frac{d}{d\tau}[f(\tau)\phi(\tau)]\right]_{\tau=0} = -f'(0)\phi(0) - f(0)\phi'(0).$$

This gives the formal equivalence

$$\phi(t)\delta'(t) \equiv \phi(0)\delta'(t) - \phi'(0)\delta(t). \tag{3.24}$$

In the same way, if ϕ is a fixed function which is twice continuously differentiable in a neighbourhood of the origin, then we should get

$$\int_{-\infty}^{+\infty} f(\tau)[\phi(\tau)\delta^{(2)}(\tau)]d\tau = \int_{-\infty}^{+\infty} [f(\tau)\phi(\tau)]\delta^{(2)}(\tau)d\tau$$

$$= \left[\frac{d^2}{d\tau^2}[f(\tau)\phi(\tau)]\right]_{\tau=0}$$

$$= f^{(2)}(0)\phi(0) + 2f'(0)\phi'(0) + f(0)\phi^{(2)}(0)$$

so that

$$\phi(t)\delta^{(2)}(t) \equiv \phi(0)\delta^{(2)}(t) - 2\phi'(0)\delta'(t) + \phi^{(2)}(0)\delta(t). \tag{3.25}$$

In general, if ϕ has continuous derivatives at least up to order n, then $\phi(t)\delta^{(n)}(t)$ reduces to

$$\phi(0)\delta^{(n)}(t) - n\phi'(0)\delta^{(n-1)}(t) + \frac{n(n-1)}{2!}\phi^{(2)}(0)\delta^{(n-2)}(t) - \cdots$$

$$\cdots + (-1)^n\phi^{(n)}(0)\delta(t). \tag{3.26}$$

Example. The expressions $\sin(t)\delta'(t)$ and $\cos(t)\delta'(t)$ may be replaced by simpler equivalent expressions as follows:

$$\sin(t)\delta'(t) = (\sin(0))\delta'(t) - (\cos(0))\delta(t) = -\delta(t),$$

$$\cos(t)\delta'(t) = (\cos(0))\delta'(t) - (-\sin(0))\delta(t) = \delta'(t).$$

This can be confirmed by examining typical integrals containing the products concerned and then reducing them to simpler equivalent forms. Thus, if $f(t)$ is an arbitrary continuously differentiable function, we have

$$\int_{-\infty}^{+\infty} f(\tau)[\sin(\tau)\delta'(\tau)]d\tau = \int_{-\infty}^{+\infty} [f(\tau)\sin(\tau)]\delta'(\tau)d\tau$$

$$= \left[-\frac{d}{d\tau}(f(\tau)\sin(\tau))\right]_{\tau=0} = -f(0) = -\int_{-\infty}^{+\infty} f(\tau)\delta(\tau)d\tau$$

which shows that $\sin(t)\delta'(t)$ is operationally equivalent to $-\delta(t)$. In the same way for $\cos(t)\delta'(t)$ we have

$$\int_{-\infty}^{+\infty} f(\tau)[\cos(\tau)\delta'(\tau)]d\tau = \int_{-\infty}^{+\infty} [f(\tau)\cos(\tau)]\delta'(\tau)d\tau$$

$$= \left[-\frac{d}{d\tau}(f(\tau)\cos(\tau))\right]_{\tau=0} = -f'(0) = \int_{-\infty}^{+\infty} f(\tau)\delta'(\tau)d\tau.$$

3.5 **POINTWISE DESCRIPTION OF** $\delta'(t)$

3.5.1 In deriving the foregoing results, no appeal has been made to any supposed pointwise behaviour of the derivatives δ', $\delta^{(2)}$, ...,, etc. As in the case of the delta function itself no such appeal is actually necessary and it is usually better avoided. Moreover, as the following considerations show, the apparent behaviour of even the first derivative, δ', at the origin is extremely difficult to describe in terms of ordinary functions.

Consider the sequence (d_n) of functions chosen in Sec.2.3 of Chapter 2 to exhibit the delta function, δ, as a limit. For each n the (generalised) derivative Dd_n consists of a pair of delta functions located at the points $-1/2n$, $+1/2n$. These are shown in Fig.3.1 as spikes in accordance with a well-established convention. As n increases so the delta functions increase in strength and approach nearer and nearer to the origin from either side. In the limit we apparently have to deal with coincident delta functions of opposite sign and of arbitrarily large strength, and it is far from easy to understand what seems to be going on.

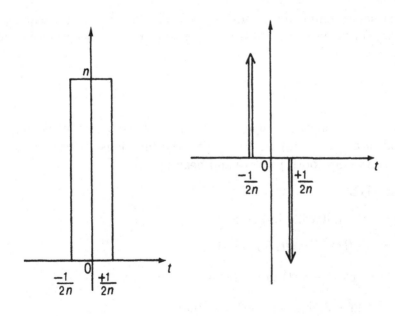

Figure 3.1:

However, once we abandon the attempt to devise a pointwise interpretation of $\delta'(t)$, there is no real problem to be considered. The significance of the sequence (Dd_n) is quite easy to interpret in a purely operational sense. Thus, suppose that f is any function which is continuously differentiable on a neighbourhood of $t = 0$. Then, for all sufficiently large values of n,

$$\int_{-\infty}^{+\infty} f(\tau)Dd_n(\tau)d\tau$$

$$= \int_{-\infty}^{+\infty} f(\tau)n\left\{\delta(\tau + \frac{1}{2n}) - \delta(\tau - \frac{1}{2n})\right\}d\tau$$

$$= n\int_{-\infty}^{+\infty} f(\tau)\delta(\tau + \frac{1}{2n})d\tau - n\int_{-\infty}^{+\infty} f(\tau)\delta(\tau - \frac{1}{2n})d\tau$$

$$= \frac{f(-1/2n) - f(1/2n)}{(1/n)}$$

and, as n goes to infinity, this tends to the limit $-f'(0)$.

3.5.2 The same point can be made, less dramatically, by using a sequence of ordinary functions which converges (in the operational sense) to δ'. For example, if we write

$$p_n(t) = \frac{n^2}{4}[u(t + 2/n) - 2u(t) + u(t - 2/n)]$$

then δ' can be taken as the (operational) limit of the $p_n(t)$. In the neighbourhood of $t = 0$, as n goes to infinity, the functions p_n assume arbitrarily large values, both positive and negative.

Exercises III.

1. Evaluate the following integrals:

(a) $\int_{-\infty}^{+\infty} e^{at}\sin(bt)\delta^{(n)}(t)dt$, for $n = 1, 2, 3$;

(b) $\int_{-\infty}^{+\infty}[\delta(t)\cos(t) - \delta'(t)\cos(2t)].[1 + t^2]^{-1}dt$;

(c) $\int_{-\infty}^{+\infty}[t^3 + 2t + 3][\delta'(t - 1) + \delta''(t - 2)]dt$.

2. Show that δ' behaves like an odd function, that is, that the expression $\delta'(-t)$ is formally equivalent to the expression $-\delta'(t)$.

3. Show that if ϕ is a twice continuously differentiable function then,

$$\frac{d}{dt}[\phi(t)\delta'(t)] = \phi'(t)\delta'(t) + \phi(t)\delta''(t)$$

in the sense that both sides have the same operational significance.

3.6 INTEGRATION OF THE DELTA FUNCTION

3.6.1 The significance of the delta function (and that of each of its derivatives) is intimately bound up with a certain conventional use of the notation for integration. This convention is consistent with the classical role of integration as a process which is, in some sense, inverse to that of differentiation. Thus, recall the statement made in Sec. 2.2.1 that the delta function, δ, is to be regarded as the derivative of the unit step function u. As already remarked in Sec. 2.2.3 this statement is to be interpreted in the following sense;

If $t \neq 0$ the function $u(t-\tau)$ is a continuous function of τ in a neighbourhood of $\tau = 0$. Accordingly the formal product $u(t-\tau)\delta(\tau)$ is meaningful and we may write

$$\int_{-\infty}^{t} \delta(\tau)d\tau = \int_{-\infty}^{+\infty} u(t-\tau)\delta(\tau)d\tau = u(t) \qquad (3.27)$$

In particular this gives

$$\int_{-\infty}^{+\infty} \delta(\tau)d\tau = 1. \qquad (3.28)$$

More generally, recall that we define $\delta^{(n)}$ as the derivative of $\delta^{(n-1)}$:

$$\delta^{(n)}(t) = \frac{d}{dt}\delta^{(n-1)}(t),$$

and it is consistent with the definitions of $\delta^{(n)}$ and $\delta^{(n-1)}$ to write

$$\int_{-\infty}^{t} \delta^{(n)}(\tau)d\tau = \int_{-\infty}^{+\infty} u(t-\tau)\delta^{(n)}(\tau)d\tau = \delta^{(n-1)}(t), \qquad n \geq 1. \quad (3.29)$$

Finally, since the function $f(t) \equiv 1$ is, trivially, infinitely differentiable we have the result

$$\int_{-\infty}^{+\infty} \delta^{(n)}(\tau)d\tau = 0 \quad \text{for } n \geq 1. \tag{3.30}$$

3.6.2 Finite limits. In general for definite integrals involving one or more finite limits of integration the presence of delta functions (or of derivatives of delta functions) in the integrand can present problems of interpretation. The fundamental sampling property which characterises, say, $\delta^{(n)}$ presupposes continuity of the nth derivative at least on a neighbourhood of the origin. If the range of integration includes the origin as an interior point (i.e. if the integral is taken over $[a, b]$, where $a < 0 < b$) then there is no difficulty: the function $u(t-a) - u(t-b)$ is certainly continuous and differentiable to all orders on a neighbourhood of the origin and we have

$$\int_{a}^{b} f(\tau)\delta^{(n)}(\tau)d\tau \equiv \int_{-\infty}^{+\infty} f(\tau)[u(\tau - a) - u(\tau - b)]\delta^{(n)}(\tau)d\tau$$

$$= (-1)^n f^{(n)}(0), \quad (n \geq 0). \tag{3.31}$$

Again, if the origin lies wholly outside the interval $[a, b]$ (so that $0 < a$, or $b < 0$) then no problem arises; trivially we have

$$\int_{a}^{b} f(\tau)\delta^{(n)}(\tau)d\tau = \int_{-\infty}^{+\infty} f(\tau)[u(t - a) - u(t - b)]\delta^{(n)}(\tau)d\tau = 0. \tag{3.32}$$

(All these remarks obviously apply to any integrand containing a factor of the form $\delta_c^{(n)}(t)$ if we replace 'the origin' by 'the point c' throughout.)

The trouble starts whenever the origin coincides with one or other of the limits of integration. Specifically, consider the case of the expression

$$\int_{0}^{+\infty} \delta(\tau)d\tau \equiv \int_{-\infty}^{+\infty} \delta(\tau)u(\tau)d\tau. \tag{3.33}$$

At first sight it is tempting to accept the suggestion, frequently made, that this integral should be assigned the value $1/2$, since "the delta function is known to behave like an even function". But we have already seen, in Sec. 3.2.3, that the formal product $u(t)\delta(t)$ is undefined, and that the most that can be said is that it should be regarded as equivalent to $k\delta(t)$, where k is some arbitrary constant. It is true that there are some other reasons for choosing the value $1/2$, but the grounds for this choice are rather more questionable than may at first appear. It is important to appreciate that there is no unique value for k which satisfies all requirements. The issue cannot be resolved by any appeal to a fundamental definition of the delta function as the limit of an appropriately chosen delta sequence. Given any specific value for k it is easy to choose a delta sequence (p_n) such that for any test function f we have

$$\lim_{n \to \infty} u(t)p_n(t)f(t)dt = kf(0).$$

Nor is the formulation of $\delta(t)$ in terms of a (Riemann) Stieltjes integral any better. In this case, we would have for an arbitrary continuous integrand f

$$\int_0^{+\infty} f(t)du_c(t) = (1-c)f(0).$$

(see No.4 of Exercises III, Chapter 2). Choosing $c = 1/2$ certainly assigns the value $1/2$ to the formal integral (3.33), but we could equally well choose $c = 0$, to give it the value 1, or $c = 1$ to give it the rather more disturbing value 0.

A specific choice for the value of the constant involved in the equivalence $u(t)\delta(t) \equiv k\delta(t)$ will need to be made when we consider the definition of the Laplace transform of $\delta(t)$ in Chapter 5.

3.7 CHANGE OF VARIABLE

3.7.1 A simple chain rule. Bearing the remarks of the preceding section in mind it remains to consider the problem of change of variable where integrals involving delta functions are concerned: that is, we need to be able to interpret expressions like $\delta^{(n)}[\phi(t)]$ when they occur in an integrand. Recall first that, for ordinary, well-behaved functions, the substitution $x = \phi(t)$ leads to an equivalence of the form

$$\int_a^b f(t)h[\phi(t)]dt = \int_{\phi(a)}^{\phi(b)} f[\phi^{-1}(x)]h(x)[\phi^{-1}(x)]'dx \qquad (3.35)$$

where, for $\phi(a) \leq x \leq \phi(b)$, we assume that $\phi^{-1}(x)$ is uniquely defined and has a uniquely defined derivative $[\phi^{-1}(x)]'$. By imposing certain restrictions on $\phi(t)$ we can establish the validity of (3.35) when $h(..)$ is replaced by $\delta(..)$, $\delta'(..)$, and so on.

However, it is often the case that a simple heuristic argument is enough to reduce the integrals concerned to an equivalent form which is easy to evaluate. In general when $x = \phi(t)$ we have the following formal extension of the usual **chain rule**:

$$\delta(x) = \frac{d}{dx}u(x) = \frac{d}{dt}u[\phi(t)]\frac{dt}{dx} = \frac{d}{dt}u[\phi(t)]/\frac{d\phi}{dt} \qquad (3.36)$$

and this last expression is easy to interpret whenever $u[\phi(t)]$ itself is reduced to a simpler form. Granted this much the method readily extends to delta functions of higher order, since we have

$$\delta'(x) = \frac{d}{dx}\delta(x) = \frac{d}{dt}\delta[\phi(t)]\frac{dt}{dx} = \frac{d}{dt}\delta[\phi(t)]/\frac{d\phi}{dt}$$

and so on. The process is most easily explained in terms of one or two explicit examples, as discussed below.

3.7.2 Monotone function. Suppose first that $\phi(t)$ increases monotonely from $\phi(a)$ to $\phi(b)$ as t goes from a to b: suppose also that $\phi(c) = 0$ for some point c such that $a < c < b$ and that $\phi(a) < \phi(c) < \phi(b)$. Using the fact that

$$\frac{d}{dx}[\phi^{-1}(x)] = 1/\frac{d}{dt}[\phi(t)]$$

we have

$$\int_a^b f(t)\delta[\phi(t)]dt = \int_{\phi(a)}^{\phi(b)} f[\phi^{-1}(x)]\delta(x)[\phi^{-1}(x)]'dx$$

$$= f([\phi^{-1}(0)])[\phi^{-1}(x)]'_{x=\phi(c)} = \frac{1}{\phi'(c)}f(c).$$

If $\phi(t)$ happens to decrease monotonely as t goes from a to b then the same result is obtained apart from a change of sign. Since in this case we would have $\phi'(c) < 0$ the two results can be summed up in the single formula:

if $\phi(t)$ monotone, with $\phi(c) = 0$ and $\phi'(c) \neq 0$, then

$$\delta(\phi(t)) = \frac{1}{|\phi'(c)|}\delta_c(t). \qquad (3.37)$$

In particular,

$$\delta(\alpha t - \beta) = \frac{1}{|\alpha|}\delta(t - \beta/\alpha). \tag{3.38}$$

In point of fact it is simpler to derive (3.37) by noting that under the particular hypotheses on ϕ we must have the equivalences

$$u[\phi(t)] = u(t - c), \text{ for } \phi(t) \text{ monotone increasing,}$$

$$u[\phi(t)] = u(c - t), \text{ for } \phi(t) \text{ monotone decreasing.}$$

Then (3.37) follows immediately from the chain rule (3.36), since we have

$$\frac{d}{dt}u(t - c) = \delta(t - c) \text{ , and } \frac{d}{dt}u(c - t) = -\delta(t - c).$$

Similar arguments can obviously be applied to expressions of the form $\delta'[\phi(t)]$, $\delta''(t)[\phi(t)]$, and so on.

Examples.

(i) Let $I_1 = \int_0^2 e^{4t}\delta(2t-3)dt$ so that $x = 2t-3$, $t = \frac{1}{2}(x+3)$ and $dt = \frac{1}{2}dx$. Then

$$I_1 = \int_{-3}^1 e^{2(x+3)}\delta(x)\frac{1}{2}dx = \frac{1}{2}e^6.$$

Alternatively, note that $u(2t - 3) = u(t - 3/2)$; hence

$$\frac{d}{dt}u(2t - 3) = \frac{d}{dt}u(t - 3/2) = \delta(t - 3/2).$$

Since $\frac{d}{dt}(2t - 3) = 2$ we have

$$I_1 = \int_0^2 e^{4t}\frac{1}{2}\delta(t - 3/2)dt = [e^{4t}/2]_{t=3/2} = \frac{1}{2}e^6.$$

(ii) Let $I_2 = \int_0^2 e^{4t}\delta(3 - 2t)dt$, so that $x = 3 - 2t$, $t = \frac{1}{2}(3 - x)$ and $dt = -\frac{1}{2}dx$. This time we get

$$I_2 = \int_3^{-1} e^{2(3-x)}\delta(x)(-1/2)dx = \frac{1}{2}\int_{-1}^3 e^{2(3-x)}\delta(x)dx = \frac{1}{2}e^6.$$

Using the alternative approach: $u(3 - 2t) = u(3/2 - t)$ so that

$$\frac{d}{dt}u(3 - 2t) = \frac{d}{dt}u(3/2 - t) = -\delta(t - 3/2).$$

Since $\frac{d}{dt}(3 - 2t) = -2$ this gives

$$I_2 = \int_0^2 e^{4t}(-1/2)[-\delta(t - 3/2)]dt = \frac{1}{2}e^6 \text{ as before.}$$

(iii) Let

$$I_3 = \int_{-\infty}^{+\infty} [\cos(t) + \sin(t)]\delta'(t^3 + t^2 + t)dt.$$

If $\phi(t) = t^3 + t^2 + t$ then ϕ is monotone increasing with $\phi(0) = 0$ and so

$$\delta(t^3 + t^2 + t) = \frac{1}{3t^2 + 2t + 1}\frac{d}{dt}u(t^3 + t^2 + t)$$

$$= \frac{1}{3t^2 + 2t + 1}\frac{d}{dt}u(t) = \frac{1}{3t^2 + 2t + 1}\delta(t) = \delta(t).$$

Further we have

$$\delta'(t^3 + t^2 + t) = \frac{1}{3t^2 + 2t + 1}\frac{d}{dt}\delta(t^3 + t^2 + t) = \frac{1}{3t^2 + 2t + 1}\delta'(t)$$

$$= \delta'(t) - \left[-\frac{6t + 2}{(3t^2 + 2t + 1)^2}\right]\delta(t) = \delta'(t) + 2\delta(t).$$

Hence,

$$I_3 = \int_{-\infty}^{+\infty} \{\cos(t) + \sin(t)\}(\delta'(t) + 2\delta(t))dt$$

$$= -[-\sin(t) + \cos(t)]_{t=0} + 2[\cos(t) + \sin(t)]_{t=0} = 1.$$

3.7.3 A delta function pair. Now let $\phi(t) = (t - \alpha)(t - \beta)$, where $\alpha < \beta$. Then ϕ is monotone decreasing for $-\infty < t < (\alpha + \beta)/2$, and $\phi(\alpha) = 0$; on the other hand ϕ is monotone increasing for the range $(\alpha + \beta)/2 < t < +\infty$, and $\phi(\beta) = 0$. Using the results of Sec. 3.7.1 above we get

$$|\phi'(\alpha)| = |2\alpha - (\alpha + \beta)| = \beta - \alpha,$$

$$|\phi'(\beta)| = |2\beta - (\alpha + \beta)| = \beta - \alpha, \tag{3.39}$$

so that

$$\int_{-\infty}^{(\alpha+\beta)/2} f(\tau)\delta\{\phi(\tau)\}d\tau = \frac{f(\alpha)}{\beta - \alpha}$$

and

$$\int_{(\alpha+\beta)/2}^{+\infty} f(\tau)\delta\{\phi(\tau)\}d\tau = \frac{f(\beta)}{\beta-\alpha}. \qquad (3.40)$$

Hence,

$$\int_{-\infty}^{+\infty} f(\tau)\delta\{(t-\alpha)(t-\beta)\}d\tau = \frac{1}{\beta-\alpha}\{f(\alpha)+f(\beta)\}. \qquad (3.41)$$

To obtain this result from the chain rule we first note that

$$\phi(t) > 0 \text{ if } t < \alpha \text{ and if } t > \beta,$$

$$\text{while } \phi(t) < 0 \text{ if } \alpha < t < \beta.$$

Hence,

$$u[(t-\alpha)(t-\beta)] = u(\alpha-t) + u(t-\beta)$$

and so

$$\frac{d}{dt}u\{\phi(t)\} = -\delta(\alpha-t) + \delta(t-\beta) = -\delta(t-\alpha) + \delta(t-\beta).$$

Since $\phi'(t) = 2t - (\alpha+\beta)$ it follows that

$$\delta\{\phi(t)\} = \frac{d}{dt}u\{\phi(t)\}/\frac{d\phi}{dt} = \frac{-\delta(t-\alpha)}{2t-(\alpha+\beta)} + \frac{\delta(t-\beta)}{2t-(\alpha+\beta)}$$

that is,

$$\delta\{(t-\alpha)(t-\beta)\} = \frac{1}{\beta-\alpha}\{\delta(t-\alpha)+\delta(t-\beta)\}, \quad \alpha < \beta \qquad (3.42)$$

which is equivalent to the result obtained in (3.41).
In particular, if $\beta > 0$ and $\alpha = -\beta$, then

$$\delta(t^2 - \beta^2) = \frac{1}{2\beta}[\delta(t+\beta) + \delta(t-\beta)]. \qquad (3.43)$$

3.7.4 A periodic train of delta functions. A result of particular importance is obtained when we take $\phi(t)$ to be a function like $\sin(t)$, which has infinitely many simple zeros, periodically distributed. A straightforward, but rather tedious, analysis of $\delta(\sin t)$ can be given as follows:
If m is any integer then ϕ is monotone increasing for all t such that

$$(2m - 1/2)\pi < t < (2m + 1/2)\pi$$

and $\phi(2m\pi) = 0$. Similarly ϕ is monotone decreasing for

$$(2m + 1/2)\pi < t < (2m + 3/2)\pi$$

and $\phi\{(2m + 1)\pi\} = 0$, $(m = 0, \pm1, \pm2, \ldots)$. As a result we can write

$$\int_{(2m-1/2)\pi}^{(2m+1/2)\pi} f(\tau)\delta(\sin\tau)d\tau = \frac{1}{|\cos(2m\pi)|}f(2m\pi) = f(2m\pi),$$

and

$$\int_{(2m+1/2)\pi}^{(2m+3/2)\pi} f(\tau)\delta(\sin(\tau))d\tau = \frac{1}{|\cos(2m+1)\pi|}f\{(2m+1)\pi\}$$

$$= f\{(2m+1)\pi\}.$$

Combining these expressions for all possible values of m gives

$$\int_{-\infty}^{+\infty} f(\tau)\delta(\sin(\tau))d\tau = \sum_{n=-\infty}^{+\infty} f(n\pi) \qquad (3.43)$$

provided that the series on the right-hand side converges. The expression $\delta(\sin(t))$ thus represents a generalised function which samples any continuous function $f(t)$ at each of the points $t = n\pi$ and then sums the resulting values. Hence it must be operationally equivalent to an infinite (periodic) train of delta functions located at the points $t = n\pi$ where $n = 0, \pm1, \pm2, \ldots$. Note that if f is any continuous function which vanishes identically outside some finite interval $[a, b]$ then no problem of convergence arises in (3.44). For in that case there must exist integers n_1, n_2, such that

$$(n_1 - 1)\pi \leq a < b \leq (n_2 + 1)\pi$$

and at worst we have only to deal with a finite sum

$$\int_{-\infty}^{+\infty} f(\tau)\delta(\sin(\tau))d\tau = \sum_{m=n_1}^{n_2} f(m\pi). \qquad (3.44)$$

The alternative chain rule approach to the evaluation of $\delta(\sin(t))$ is particularly instructive. As the successive diagrams in Fig. 3.2 indicate, the whole process can easily be carried out by inspection. For completeness, however, we give the requisite steps of the argument in detail as follows:

Since $\phi(t) > 0$ for all t such that $2m\pi < t < (2m+1)\pi$ and $\phi(t) < 0$ for all t such that $(2m-1)\pi < t < 2m\pi$ for $m = 0, \pm1, \pm2, \ldots$, it follows that

$$u(\sin(t)) = \sum_{m=-\infty}^{+\infty} [u\{t - 2mn\pi\} - u\{t - (2m+1)\pi\}]. \qquad (3.45)$$

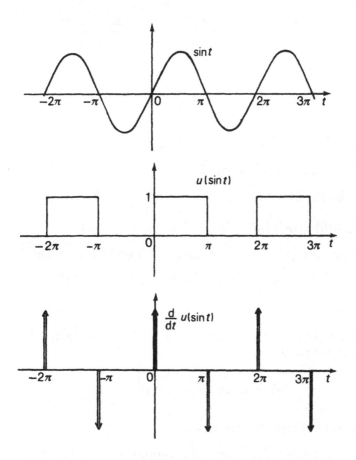

Figure 3.2:

Hence,

$$\frac{d}{dt}u(\sin(t)) = \sum_{m=-\infty}^{+\infty} [\delta\{t - 2m\pi\} - \delta\{t - (2m+1)\pi\}].$$

Now $\phi'(t) = \cos(t)$ and so, for $m = 0, \pm 1, \pm 2, \ldots$, we have

$$\cos(2m\pi) = +1 \quad \text{and} \quad \cos(2m+1)\pi = -1.$$

Thus,

$$\delta\{\sin(t)\} = \frac{1}{\cos(t)} \sum_{m=-\infty}^{+\infty} [\delta\{t - 2m\pi\} - \delta\{t - (2m+1)\pi\}]$$

$$= \sum_{m=-\infty}^{+\infty} \left[\frac{\delta\{t - 2m\pi\}}{\cos(2m\pi)} - \frac{\delta\{t - (2m+1)\pi\}}{\cos((2m+1)\pi)} \right]$$

$$= \sum_{m=-\infty}^{+\infty} \delta(t - m\pi). \tag{3.46}$$

Exercises IV.

1. Evaluate the following integrals:

(a) $\int_{-1}^{0} \sinh(2t)\delta(5t+2)dt$; (b) $\int_{-2\pi}^{+2\pi} e^{\pi t}\delta(t^2 - \pi^2)dt$;

(c) $\int_{-\pi}^{+\pi} \cosh(\theta)\delta(\cos(\theta))d\theta$; (d) $\int_{-\infty}^{+\infty} e^{-|t|}\delta(\sin(\pi t))dt$.

2. Find equivalent forms for each of the following expressions:

(a) $\delta\{\sin(|t|)\}$; (b) $\delta\{\cos(\pi t/2)\}$; (c) $\delta(e^t)$; (d) $\delta'(\theta^2 - \pi^2)$;

(e) $\delta'\{\sinh(2x)\}$.

3. If x is confined to the range $0 < x < +\infty$, find simpler equivalent forms for the expressions:

$$\text{(a)} \ \delta\{\log(x)\}; \quad \text{(b)} \ \delta\left[\frac{1}{a} - \frac{1}{x}\right] \ ;$$

where a is some fixed positive number.

4. Let ϕ be a twice continuously differentiable function which increases monotonely from $\phi(a)$ to $\phi(b)$ as t increases from a to b. If $\phi(c) = 0$ where $a < c < b$ and $\phi(a) < \phi(c) < \phi(b)$, show that

$$\delta'\{\phi(t)\} = \frac{1}{|\phi'(c)|^2}\{\delta_c'(t) + \frac{\phi''(c)}{\phi'(c)}\delta_c(t)\}.$$

Further reading.

Few of the many texts which introduce the delta function give a really thorough and and systematic account of its properties and illustrate how it should be treated in the manipulation of algebraic expressions. The lack of such explanatory detail was, indeed, one of the main motivations for the production of the present book and, in particular, for the content of this particular chapter. It is therefore not easy to recommend suitable supplementary reading other than those texts already referred to at the end of Chapter 2.

However, one such useful source of information may be found in texts on differential equations and linear differential operators. One of the most readable and readily accessible introductions to the definition and use of the delta function in this context remains the chapter on Green's functions in **Principles and Techniques of Applied Mathematics** by **Bernard Friedman**, published in the John Wiley Applied Mathematics Series as long ago as 1956.

Chapter 4

Time-invariant Linear Systems

4.1 SYSTEMS AND OPERATORS

4.1.1 Systems and signals. In this chapter we examine in some detail
one of the most natural and immediately useful applications of the generalised functions introduced in Chapters 2 and 3. The analysis of physical
systems, whether mechanical, electrical or optical, involves the mathematical modelling of the transformation and processing of measurable,
time-varying quantities. Broadly speaking this is the area which may be
described as 'signal processing'. The signals themselves admit an obvious mathematical representation in terms of functions in the orthodox
sense of the term. It is when it comes to the systems which carry out the
processing that the classical concept of function seems to be inadequate,
and when a need for generalised functions such as the delta function and
its derivatives first becomes apparent.

To begin with recall once again that a function could be defined as a rule,
or formula, which maps or transforms numbers into numbers. Similarly
an operator may be defined as a rule or formula which maps or transforms
functions into functions. By a **system** we shall therefore understand a
mathematical model of a physical device which may be represented by
(or identified with) an operator, T, which maps the members of one class
of functions ('input signals') into those of another ('output signals').

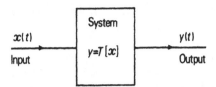

Figure 4.1:

Throughout this chapter we shall generally denote input signals by x_1, x_2, etc., and output signals by y_1, y_2, etc. The relation between a given input signal x and the corresponding output signal y may be indicated either by specific use of an operator symbol T as in

$$y = T[x] \tag{4.1}$$

or simply by writing

$$x(t) \ \rightarrow \ y(t). \tag{4.2}$$

We shall refer to y as the **image of x under** T or sometimes as the **response of the system** represented by the operator T to the input x. Where convenient we shall use the ordinary block diagram representation as shown in Fig. 4.1.

4.1.2 Linearity. An operator T is said to be **linear** whenever it enjoys the following properties:

L1. **Additivity.** If $y_1 = T[x_1]$ and $y_2 = T[x_2]$, then

$$T[x_1 + x_2] = y_1 + y_2 \equiv T[x_1] + T[x_2]. \tag{4.3}$$

L2. **Homogeneity.** If $y = T[x]$ and if α is any real number, then

$$T[\alpha x] = \alpha y \equiv \alpha\{T[x]\}. \tag{4.4}$$

A system is linear if and only if the operator T which represents it is a linear operator in the sense defined above. Hence a linear system is one for which the **Principle of Superposition** is satisfied, that is, one for which the response to any finite linear combination of inputs is always the like linear combination of outputs. Thus, if $x_1 \ \rightarrow \ y_1(t)$ and

$x_2(t) \quad \rightarrow \quad y_2(t)$, and if α and β are any real numbers, then for a linear system we have

$$\alpha x_1(t) + \beta x_2(t) \quad \rightarrow \quad \alpha y_1(t) + \beta y_2(t).$$

Note that from this definition of linearity it follows at once that if $x(t) = 0$ for all t then so also does $y(t)$: that is, there can be no output without some (non-zero) input. This means that there is no stored energy; the definition restricts us to relaxed systems.

Examples.

(i) **Constant gain amplifier.** The output y is always a constant multiple of the input x,

$$x(t) \quad \rightarrow \quad y(t) = Ax(t).$$

(ii) **Multiplier.** The output y is the product of the input x and a certain fixed function ϕ,

$$x(t) \quad \rightarrow \quad y(t) = \phi(t)x(t).$$

(iii) **Differentiator.** The output is the first derivative of the input,

$$x(t) \quad \rightarrow \quad y(t) = x'(t).$$

(iv) **Integrator.** An integrator is a linear system which carries out a transformation process inverse to that of the differentiator. That is, it is a system in which the input x turns out to be the derivative of the output, y. We shall adopt the following definition:

$$x(t) \quad \rightarrow \quad y(t) = \int_{-\infty}^{t} x(\tau)d\tau.$$

(v) **Ideal time delay.** If f is any function of t and if a is any fixed real number, recall that the **translate of f with respect to** a is the function f_a given by

$$f_a(t) = f(t - a) \quad \text{for all} \quad t.$$

If a is positive then the **ideal time delay of duration** a may be defined as the system whose response to any input x is the translate x_a:

$$x(t) \quad \rightarrow \quad x(t - a) \equiv x_a(t).$$

4.1.3 Real systems. For a model of an actual physical system (however idealised) we would naturally require that all input and output signals should be real-valued functions of time. However it is often convenient to work in terms of complex-valued signals. The concept of linearity for a real system can be extended to allow for this eventuality in the following, fairly obvious, way:

L3. **Realness.** Let the responses of the system to the real-valued inputs $x_1(t)$ and $x_2(t)$ be $y_1(t)$ and $y_2(t)$ respectively. Then the response to the complex-valued input signal $z(t) = x_1(t) + ix_2(t)$ is (defined to be) the complex-valued function $w(t) = y_1(t) + iy_2(t)$. (However, real inputs always give real outputs.)

4.1.4 Time-invariance. An operator T is said to be **stationary** if it has the following property. If $y = T[x]$ and if a is any fixed real number, then

$$T[x_a] = y_a \equiv \{T[x]\}_a. \tag{4.5}$$

Any system represented by a stationary operator T is said to be **time-invariant**. Thus, for a time-invariant system, the only effect of delaying (or advancing) an input signal is to produce a corresponding delay (or advance) of the output signal: if $x(t) \rightarrow y(t)$ then, for any a,

$$x(t-a) \rightarrow y(t-a).$$

A system which is both linear and time-invariant will often be referred to as a T.I.L.S. (time-invariant linear system). With the exception of the multiplier all the systems described above are both linear and time-invariant. In general a multiplier is a linear, **time-varying** system. By contrast, the system defined by the input/output relation

$$x(t) \rightarrow [x(t)]^2$$

is clearly time-invariant but, equally clearly, non-linear.

4.1.5 Remarks.

(i) In practice, many systems of interest are described in terms of input/output relations which take the form of differential equations or integro-differential equations. The significance of linearity (and, to some extent, of time-invariance) needs to be examined rather more closely in such cases, because of the problems associated with the specification of initial conditions. We consider this point in detail in the example discussed at the end of this chapter.

(ii) It is worth emphasising that, to begin with, all signals are considered to be functions in the usual sense of the word. The philosophy adopted here is that, in principle at least, it should always be possible to measure the amplitude of any physical signal at any given time. Hence although the use of generalised functions as inputs or outputs will prove to be convenient for analytical purposes, it is essentially an idealisation and should come logically as an extension of the basic theory.

More particularly we should expect a system to be defined initially in terms of its responses to the members of some specific class of 'admissible' inputs, in the sense that the corresponding outputs are always well-defined as functions in the classical sense. Thus, for example, only those functions which are everywhere differentiable in the classical sense would be considered to be admissible inputs for the differentiator in the first instance. To define the response of a differentiator to a discontinuous input signal (such as a unit step function) would clearly require us to extend the concept of signal and admit generalised functions as signals; the corresponding output signal in this case would, of course, be a delta function.

(iii) In general a system can be described as 'continuous' if, whenever a sequence of input signals converges to zero (in some sense to be specified), the corresponding sequence of output signals also converges to zero (again in some specific sense). In certain circumstances this is equivalent to the demand that 'small' inputs should always give rise to correspondingly 'small' outputs. We will not pursue this question further here, but from time to time we shall indicate in the text those points where the assumption of system continuity becomes important. For the moment we merely note the following fact:

Linearity is often taken to mean simply that the Principle of Superposition is valid in the sense that the image of the pointwise sum of two admissible inputs is the like pointwise sum of the individual responses. Strictly, this only says that the system is additive (L1). However, whenever an appropriate continuity condition is imposed, every additive system turns out to be automatically homogeneous and therefore linear in the full sense of the word.

Exercises I.

1. Establish which of the systems defined below in terms of input/output relations is (i) linear, and (ii) time-invariant.

(a) $x(t) \longrightarrow x''(t) - 2x(t)$;

(b) $x(t) \longrightarrow \int_0^t x(\tau)d\tau + x(t)\sin(t)$;

(c) $x(t) \longrightarrow kx(0)$, where k is a real, non-zero, constant;

(d) $x(t) \longrightarrow |x(t)|$;

(e) $x(t) \longrightarrow x_+(t) \equiv \max[x(t), 0]$.

2. If T is an additive operator, prove that

(i) for every (admissible) function x we have

$$T[-x] = -T[x].$$

(ii) T is always homogeneous at least with respect to rational multipliers; that is to say,
$$T[rx] = rT[x]$$
for every function x and every rational number r.

3*. Let $h(t, \tau)$ be a bounded, continuous function of τ for each fixed t. If x and y are related by

$$y(t) = \int_{-\infty}^{+\infty} x(\tau)h(t, \tau)d\tau,$$

show that the mapping $x(t) \longrightarrow y(t)$ defines a system which is linear (but not necessarily time-invariant). If the system does happen to be time-invariant, show that the function $h(t, \tau)$ actually takes the specific form

$$h(t, \tau) \equiv h(t - \tau).$$

4.2 STEP RESPONSE AND IMPULSE RESPONSE

4.2.1 Definition of impulse response. Given a T.I.L.S. represented by an operator T, suppose that the unit step function, u, is an admissible

input. That is to say, assume that in response to the unit step applied
as an input there exists a well-defined function $\sigma = T[u]$. The function
sketched in the second graph of Fig. 4.2 is intended to illustrate just such
a response to the unit step function input shown in the first graph. This
function σ is usually referred to as the **step-response** of the system.

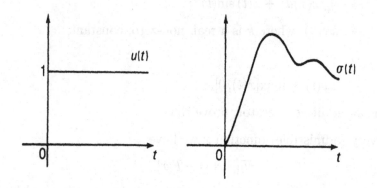

Figure 4.2: Unit step input and step response

Now consider the system response to the function

$$x(t) = k[u(t + a/2) - u(t - a/2)] \qquad (4.6)$$

which represents a rectangular pulse of duration a and of amplitude k,
centred about the origin. Since the system concerned is both linear and
time-invariant the response to this signal applied as an input can be
expressed as a linear combination of suitably translated versions of the
step-response. (See Fig. 4.3 for a sketch of the rectangular pulse input,
expressed as a combination of step functions, and of the corresponding
output.)

That is to say we will have

$$y(t) \equiv T[x](t) = T[k\{u(t + a/2) - u(t - a/2)\}]$$
$$= k\{T[u(t + a/2)] - T[u(t - a/2)]\}$$
$$= k\{\sigma(t + a/2) - \sigma(t - a/2)\}. \qquad (4.7)$$

Now suppose further that the function σ has a continuous classical derivative, $h = \sigma'$. If the pulse width a is sufficiently small, then we can write

$$k\{\sigma(t + a/2) - \sigma(t - a/2)\} \simeq (ka)h(t) \qquad (4.8)$$

(For, if σ is continuously differentiable, then the first mean value theorem of the differential calculus gives

$$\sigma(t + a/2) - \sigma(t - a/2) = \sigma'(t + \xi)$$

where ξ is some point such that $-a/2 < \xi < +a/2$. If a is small then by the continuity of σ we have

$$\sigma'(t + \xi) = \sigma'(t) + \epsilon \simeq \sigma'(t).)$$

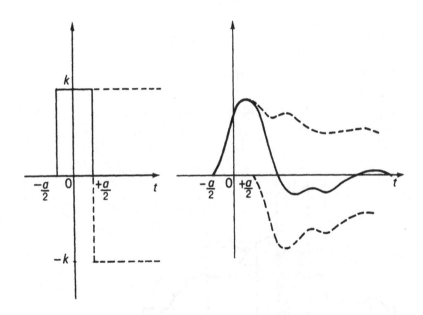

Figure 4.3:

Taking the particular case when $k = 1/a$ we can say that the function h, defined here as the derivative of the step response σ, represents to a first approximation the system response to a narrow pulse of unit area located at the origin. This function h is called the **impulse response** of the system.

4.2.2 Convolution integral. Now take for the input x a bounded, continuous function which vanishes identically outside some finite time-interval, say $\alpha \le t \le \beta$. We can approximate $x(t)$ by a train of adjacent narrow pulses as illustrated in Fig. 4.4,

$$x(t) \simeq \sum_k x(\tau_k)[u(t - t_k) - u(t - t_{k-1})].$$

The response of the system to an elementary pulse of width $\Delta\tau_k$, where $\Delta\tau_k = t_k - t_{k-1}$, and of height $x(\tau_k)$, centred about the point $t = \tau_k$, will be given (approximately) by

$$\{x(\tau_k)\Delta\tau_k\}h(t - \tau_k),$$

using (4.8), and the fact that the system is time-invariant. Next, by linearity of the system, the response to the input $x(t)$ will be given, again to a first approximation, by the sum

$$\sum_k x(\tau_k)h(t - \tau_k)\Delta\tau_k. \qquad (4.9)$$

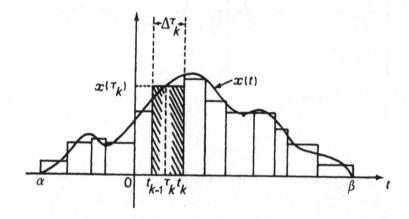

Figure 4.4:

In the limit, as the approximating pulses are made narrower and narrower, we obtain the actual output y in the form of a so-called **convolution integral**

$$y(t) = \int_\alpha^\beta x(\tau)h(t - \tau)d\tau = \int_{-\infty}^{+\infty} x(\tau)h(t - \tau)d\tau. \qquad (4.10)$$

We can take infinite limits in (4.10) because the function x is known to vanish identically outside the finite interval $[\alpha, \beta]$. If we remove this constraint on x then an extension of the above argument can be used to show that the output y will still be given by an integral of the form (4.10), provided that the functions x and h are suitably well-behaved for large values of $|t|$ (so that the infinite integral does converge). Similarly we may weaken considerably the assumption that x and h are continuous. Note, however, that the one crucial step in the argument is the passage to the limit in the derivation of (4.10) from (4.9). It is this step which relies explicitly on a suitable assumption of system continuity as discussed in the Remarks at the end of section 4.1 above.

4.3 CONVOLUTION

4.3.1 Algebraic properties of convolution. By a simple change of variable we have

$$y(t) = \int_{-\infty}^{+\infty} x(\tau)h(t-\tau)d\tau = \int_{-\infty}^{+\infty} x(t-\tau)h(\tau)d\tau. \qquad (4.11)$$

One immediate consequence of this commutative property of the convolution integral is that the input function x and the impulse response function h may be regarded as interchangeable, in a certain sense. We could equally well regard y as the response of a system whose impulse response function is x to the input signal h.

Now consider two T.I.L.S., with impulse response functions h_1 and h_2 respectively, connected in cascade (Fig. 4.5). The response z to a continuous input x can be computed as follows (always assuming that the change in order of integration can be justified).

$$z(t) = \int_{-\infty}^{+\infty} y(t-\theta)h_2(\theta)d\theta$$

$$= \int_{-\infty}^{+\infty} h_2(\theta) \left[\int_{-\infty}^{+\infty} x(t-\theta-\phi)h_1(\phi)d\phi \right] d\theta$$

$$= \int_{-\infty}^{+\infty} h_2(\theta) \left[\int_{-\infty}^{+\infty} x(t-\tau)h_1(\tau-\theta)d\tau \right] d\theta,$$

putting $\tau = \theta + \phi$,

$$= \int_{-\infty}^{+\infty} x(t - \tau) h_3(\tau) d\tau,$$

where $h_3(\tau) = \int_{-\infty}^{+\infty} h_1(\tau - \theta) h_2(\theta) d\theta$. Thus the cascade connection of the two T.I.L.S. is equivalent to a single T.I.L.S. whose impulse response function is the convolution of the individual impulse functions of the connected systems.

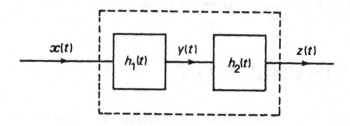

Figure 4.5:

4.3.2 Notation for convolution. As an operation on functions it is usual to denote convolution by means of the symbol $\star$:

$$h_3(t) = h_1(t) \star h_2(t) \equiv \int_{-\infty}^{+\infty} h_1(t - \tau) h_2(\tau) d\tau. \qquad (4.12)$$

It can be shown that whenever the functions h_1 and h_2 are each absolutely integrable over $(-\infty, +\infty)$ then the convolution is well-defined and h_3 is itself a function absolutely integrable over $(-\infty, +\infty)$. Furthermore, under these conditions, the derivation of the results of section 4.3.1 is readily justified. These results show that the operation is both commutative and associative:

$$(h_1 \star h_2)(t) = (h_2 \star h_1)(t)$$

$$[h_1 \star (h_2 \star h_3)](t) = [(h_1 \star h_2) \star h_3](t).$$

Likewise it is easy to confirm that the distributive law holds:

$$[(h_1 + h_2) \star h_3](t) = (h_1 \star h_3)(t) + (h_2 \star h_3)(t).$$

Thus, from a purely formal point of view, the operation of forming the convolution of functions has much in common with multiplication.

Exercises II.

1. Find the step response and the impulse response of the ideal integrator, defined by the input/output relation $y(t) = \int_{-\infty}^{t} x(\tau)d\tau$.

2. Evaluate each of the following convolutions:

(a) $u(t) \star u(t)$; (b) $e^{\alpha t} \star [u(t)\cos(\omega t)]$, $\alpha > 0$, $\omega \neq 0$;

(c) $[u(t)\sin(t)] \star [u(t)\cos(t)]$; (d) $p(t) \star p(t)$,

where the function p is given by $p(t) = u(t + 1/2) - u(t - 1/2)$.

3. Let f be a continuous function which vanishes for all $t < 0$ and define functions g_n as follows:

$$g_n(t) = \int_0^t \frac{(t-\tau)^n}{n!} f(\tau)d\tau, \quad \text{for} \ n \geq 1; \quad g_0(t) = \int_0^t f(\tau)d\tau.$$

Show that g_n is the result of integrating f $(n+1)$ times from 0 to t. [Hint: use the Leibnitz formula for differentiating an integral with respect to a parameter.]

4. Let f be an arbitrary continuous function which vanishes for all $t < 0$. Then if $g(t) = u(t)/\sqrt{\pi t}$, prove that $f(t) \star g(t)$ is the 'half-integral' of f from 0 to t, in the sense that

$$[f(t) \star g(t)] \star g(t) = \int_0^t f(\tau)d\tau.$$

4.4 IMPULSE RESPONSE FUNCTIONS

4.4.1 The delta function as an impulse response function. Consider in particular the case of the time-invariant linear system which has the property of leaving every input unchanged. If we assume that there exists a convolution integral representation for this system, then the analogue of equation (4.10) should take the form

$$x(t) = \int_{-\infty}^{+\infty} x(t-\tau)h(\tau)d\tau = \int_{-\infty}^{+\infty} x(\tau)h(t-\tau)d\tau \qquad (4.13)$$

where h denotes the impulse response function characterising the system, and x is an arbitrary continuous function. Putting $t = 0$ in (4.13) we get

$$x(0) = \int_{-\infty}^{+\infty} x(-\tau)h(\tau)d\tau = \int_{-\infty}^{+\infty} x(\tau)h(-\tau)d\tau. \qquad (4.14)$$

Now write $z(t) \equiv x(-t)$, and apply (4.14) directly to the function z in place of x:

$$z(0) = \int_{-\infty}^{+\infty} z(-\tau)h(\tau)d\tau \equiv \int_{-\infty}^{+\infty} x(\tau)h(\tau)d\tau. \qquad (4.15)$$

But, since $z(0) = x(0)$, this last equation is simply a restatement of the fundamental sampling property of the delta function, and it follows that the system in question must have the delta function itself as its impulse response function. (This is, of course, consistent with the fact that the unit step function, applied as an input, would be transmitted unchanged: that is, that the step response of the system is the unit step function.)

Hence, as remarked at the beginning of this chapter, we shall find it necessary to introduce the concept of generalised function in order to obtain a mathematical model for what is, after all, an extremely simple and basic time-invariant linear system - namely, for an ideal amplifier of unit gain. From the mathematical point of view this introduction of the delta function is important in another sense. The parallel between convolution as an operation on functions with ordinary pointwise multiplication has already been observed in so far as the associative, commutative and distributive laws are concerned. The delta function is now seen to play the role of a unit element with respect to convolution, since we can write (4.13) in the form

$$x(t) = (x \star \delta)(t) = (\delta \star x)(t).$$

More generally the impulse response function associated with an ideal time delay of duration a is seen to be the shifted delta function, δ_a:

$$x(t - a) = \int_{-\infty}^{+\infty} x(t - \tau)\delta_a(\tau)d\tau = x(t) \star \delta_a(t). \qquad (4.16)$$

4.4.2 Impulse response of a differentiator. A similar situation arises if we seek a mathematical representation of the ideal differentiator in terms of a (formal) convolution integral. If a continuously differentiable input function x is applied to this system, then a formal appeal to equation (4.10) yields the result

$$x'(t) = \int_{-\infty}^{+\infty} x(t - \tau)h(\tau)d\tau = \int_{-\infty}^{+\infty} x(\tau)h(t - \tau)d\tau. \qquad (4.17)$$

Thus, putting $t = 0$, we have

$$x'(0) = \int_{-\infty}^{+\infty} x(-\tau)h(\tau)d\tau = \int_{-\infty}^{+\infty} x(\tau)h(-\tau)d\tau. \qquad (4.18)$$

Writing $z(t) \equiv x(-t)$, as before, and applying (4.18) to the function z,

$$z'(0) = \int_{-\infty}^{+\infty} z(-\tau)h(\tau)d\tau \equiv \int_{-\infty}^{+\infty} x(\tau)h(\tau)d\tau. \qquad (4.19)$$

This time we must take into account the fact that

$$\frac{d}{dt}z(t) = \frac{d}{dt}x(-t) = -\frac{d}{d(-t)}x(-t)$$

so that, in particular, $z'(0) = -x'(0)$, and (4.19) is seen to be the characteristic sampling property associated with the first derivative, δ', of the delta function.

In the same way, the generalised function $\delta^{(n)}$ may be interpreted as the impulse response function characterising the time-invariant linear system which transforms each n-times continuously differentiable function x into its nth derivative $x^{(n)}$.

4.4.3 Convolution of delta functions. These results lead naturally to the extension of the operation of convolution to include generalised functions. In Sec. 4.3 we have seen that the convolution of two ordinary functions, $h_1 \star h_2$, can be interpreted in terms of the cascade connection of the time-invariant linear systems of which h_1 and h_2 are the impulse response functions. If we replace one of those functions by a delta function, then the convolution $h_1 \star \delta_a$ (where h_1 is assumed to be continuous) can be interpreted as either:

(i) the impulse response function corresponding to the cascade connection of the system with impulse response function h_1 and a time delay of duration a, or

(ii) the response of a time delay of duration a to an input signal h_1, or

(iii) the response of a system with impulse response function h_1 to the delta function δ_a applied as an input.

The last of these alternatives is the one most often encountered in the classical literature of linear systems theory, and is, indeed, the genesis of the name 'impulse response' for the function h which we have defined here as the derivative of the step response.

As for the convolution of two delta functions, say $\delta_a \star \delta_b$, note that the cascade connection of two time delays of duration a and b respectively ought to result in a single time delay of duration $(a + b)$. That is to say, we should have

$$(\delta_a \star \delta_b)(t) = \int_{-\infty}^{+\infty} \delta_a(t - \tau)\delta_b(\tau)d\tau = \delta_a(t - b) = \delta_{a+b}(t). \qquad (4.20)$$

This is consistent with the following formal manipulations:

$$\int_{-\infty}^{+\infty} \delta_b(\theta) \left[\int_{-\infty}^{+\infty} f(t - \theta - \phi)\delta_a(\phi)d\phi \right] d\theta$$

$$= \int_{-\infty}^{+\infty} \delta_b(\theta) f(t - \theta - a)d\theta = f(t - b - a).$$

With regard to the derivatives of the delta function, let f be a function which has continuous derivatives up to at least the $(m+n)^{th}$ order. Then we have

$$[f(t) \star \delta^{(n)}(t)] \star \delta^{(m)}(t) = f^{(n)}(t) \star \delta^{(m)}(t) = f^{(n+m)}(t).$$

Assuming that the convolution product remains associative under these conditions it follows that, for any $m \geq 0$ and any $n \geq 0$,

$$\delta^{(n)}(t) \star \delta^{(m)}(t) = \int_{-\infty}^{+\infty} \delta^{(n)}(t - \tau)\delta^{(m)}(\tau)d\tau = \delta^{(n+m)}(t) \qquad (4.21)$$

4.5   TRANSFER FUNCTION

4.5.1 Definition of transfer function. Let $H_s(t)$ denote the response of a given T.I.L.S. to the input $x(t) = e^{st}$:

$$e^{st} \rightarrow H_s(t).$$

Since the system is time-invariant we know that, for any given fixed value of τ, the response to the input $x(t + \tau)$ must be $H_s(t + \tau)$. But since $x(t) = e^{st}$ we also know that

$$x(t + \tau) = e^{s(t+\tau)} = e^{s\tau} e^{st} = e^{s\tau} x(t).$$

Hence, since the system is linear, the response to the input $x(t+\tau)$ must be $e^{s\tau} H_s(t)$ and so we must have

$$H_s(t + \tau) = e^{s\tau} H_s(t).$$

In particular we can put $t = 0$ to get

$$H_s(\tau) = e^{s\tau} H_s(0). \tag{4.22}$$

The relation (4.22) holds for any given value of τ. Writing t (a variable) in place of τ, and denoting the number $H_s(0)$ by the symbol $H(s)$ (since it depends only on the value of the parameter s) we get the following result

The response of a (relaxed) T.I.L.S. to an exponential input, $x(t) = e^{st}$, is always of the form

$$y(t) = H(s)e^{st} \tag{4.23}$$

where $H(s)$ is a number which depends only on the particular system concerned and on the parameter s.

Allowing the parameter s in (4.23) to take complex values, $s = \sigma + i\omega$, we refer to $H(s)$ as the **transfer function** of the system concerned. Note that for a given system there will generally be inputs of the form e^{st} for which there exists no well-defined response. Equivalently, the transfer function $H(s)$ will be defined in general only for a limited range of values of s (that is, only on some region of the complex plane).

4.5.2 Frequency response and Fourier transform. Now suppose that the T.I.L.S. is one for which there exists a well-defined impulse response function $h(t)$. Using the convolution integral (4.10), the response of the system to the exponential input e^{st} is given by

$$\int_{-\infty}^{+\infty} e^{s(t-\tau)} h(\tau) d\tau = e^{st} \int_{-\infty}^{+\infty} e^{-s\tau} h(\tau) d\tau. \tag{4.24}$$

Comparing (4.23) and (4.24) we obtain the following explicit representation for the transfer function, $H(s)$:

$$H(s) = \int_{-\infty}^{+\infty} e^{-s\tau} h(\tau) d\tau \qquad (4.25)$$

whenever the infinite integral exists.

Equation (4.25) defines what is usually described as the **two-sided Laplace Transform** of the function h. If we put $s = i\omega$ then the exponential input takes the form

$$e^{i\omega t} = \cos(\omega t) + i\sin(\omega t),$$

a complex combination of real sinusoids of frequency ω. The output is the function $e^{i\omega t} H(i\omega)$ where, again provided that the infinite integral exists, we have

$$H(i\omega) = \int_{-\infty}^{+\infty} e^{-i\omega \tau} h(\tau) d\tau. \qquad (4.26)$$

The function $H(i\omega)$ is called the **frequency response function** of the system whose impulse response function is $h(t)$, or the **Fourier Transform** of the function h itself.

In most modern texts it is usual to define the **one-sided Laplace Transform** of the function h as the following:

$$\mathcal{L}[h(t)] \equiv \int_{0}^{+\infty} e^{-s\tau} h(\tau) d\tau. \qquad (4.27)$$

Note that this may be interpreted as the two-sided transform of the function $u(t)h(t)$:

$$\mathcal{L}[h(t)] = \int_{0}^{+\infty} e^{-s\tau} h(\tau) d\tau = \int_{-\infty}^{+\infty} e^{-s\tau} [u(\tau)h(\tau)] d\tau. \qquad (4.28)$$

In the case of a system whose impulse response function h vanishes for all negative values of t we have

$$y(t) = \int_{-\infty}^{t} x(\tau)h(t-\tau) d\tau = \int_{0}^{+\infty} x(t-\tau)h(\tau) d\tau \qquad (4.29)$$

and it follows at once that if the input x vanishes for all $t < 0$ then so also does the output y. A system of this kind is said to be **causal**, or

non-anticipative, in the sense that no output signal can ever precede the input signal which gives rise to it. Accordingly the one-sided Laplace Transform of an arbitrary function h could always be interpreted as the transfer function of the causal system whose impulse response function is $u(t)h(t)$. Note also that, for a causal system, if the input signal x is such that $x(t) = 0$ for all $t > 0$ then the output signal y is given by the following expression:

$$y(t) = \int_0^t x(\tau)h(t - \tau)d\tau = \int_0^t x(t - \tau)h(\tau)d\tau. \qquad (4.30)$$

(In some texts (4.30) is actually used as the definition of the convolution of two arbitrary functions x and h, whether those functions vanish for negative values of t or not. We shall not adopt this convention here.)

4.5.3 The Fourier and Laplace transforms of delta functions.

The remarks in Sec. 4.4 on the appearance of generalised functions as impulse responses suggests that Laplace and Fourier Transforms for delta functions and derivatives of delta functions could be defined in terms of the appropriate transfer functions. So far as the two-sided Laplace Transform and the Fourier Transform are concerned there is, of course, no particular merit in this since we may appeal directly to the characteristic sampling properties in each case:

$$\int_{-\infty}^{+\infty} e^{-st}\delta(t)dt = [e^{-st}]_{t=0} = 1,$$

$$\int_{-\infty}^{+\infty} e^{-st}\delta'(t)dt = -\left[\frac{d}{dt}(e^{-st})\right]_{t=0} = s,$$

and so on. In the case of the one-sided Laplace Transform, however, we are confronted with expressions of the form

$$\int_0^{+\infty} e^{-st}\delta(t)dt \equiv \int_{-\infty}^{+\infty} e^{-st}u(t)\delta(t)dt$$

and

$$\int_0^{+\infty} e^{-st}\delta'(t)dt \equiv \int_{-\infty}^{+\infty} e^{-st}u(t)\delta'(t)dt.$$

As already discussed in Section 3.6.2, these expressions are undefined in the formal calculus of delta functions since the sampling operations are required to take place at a point of discontinuity, and we are unable to offer a consistent interpretation for the product $u(t)\delta(t)$. However the system-theoretic context explored above does suggest an ad hoc solution:

The Laplace Transform of the delta function could be defined as the transfer function of the causal T.I.L.S. which transmits every input unchanged. In this sense we would be entitled to write

$$\mathcal{L}[\delta(t)] = 1.$$

Similarly, treating δ' as the impulse response of the ideal differentiator and noting that this is also most certainly a causal system, we should have

$$\mathcal{L}[\delta'(t)] = s.$$

In the following chapter we will be considering the Laplace Transform and its properties in some detail, and it will be seen that these suggested definitions are, in fact, consistent with the general theory.

Exercises III.

1. Evaluate the integral

$$\int_{-\infty}^{+\infty} [x^4 - 10x^2 + 1][\delta_{\sqrt{2}}(x) \star \delta_{\sqrt{3}}(x)]dx.$$

2. Find the frequency response functions associated with the following time-invariant linear systems:

(a) an ideal amplifier of constant gain,

(b) an ideal time delay,

(c) the ideal differentiator,

(d) the ideal integrator (as defined in Ex.II, No.1.).

3. A linear, time-invariant system has step response function $\sigma(t) = u(t)\sin(\alpha t)$. Find the output of the system if the input is the function $x(t) = u(t)\sin(\alpha t)$. What is the frequency response function of the system?

4.5.4 Example: an electrical network. The interpretation of the concepts of transfer function, step response, and impulse response can be readily illustrated in the context of electrical networks. The simple RC electrical circuit of Fig. 4.6 can be described initially in terms of the first-order differential equation

$$v(t) = R\frac{dq}{dt} + \frac{1}{C}q(t) \qquad (4.31)$$

where q(t) denotes the charge on the capacitor at time t and $v(t)$ the voltage impressed on the circuit by an independent external generator.

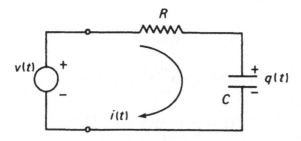

Figure 4.6:

We can write this equation in the form

$$v = Aq \qquad (4.32)$$

where $A \equiv R\frac{d}{dt} + \frac{1}{C}$. This symbol A is an operator in that it represents a transformation which carries the function q into the function v: as is easily confirmed this operator is both linear and stationary. However, this operator formulation of equation (4.31) does not correspond to the way in which the physical situation is understood when we think of it in terms of a system. Equation (4.32) exhibits the voltage v as the result of operating on the charge q. We would more commonly think of the generator as supplying a known voltage wave-form v and seek to find the resulting charge q. In brief, the system which the electrical circuit of Fig. 4.6 represents is usually thought of as that which is described by the solution of the differential equation (4.31).

Now equation (4.31) is a linear first-order equation with integrating factor $\exp(t/CR)$. Its explicit solution presents no particular difficulty, but it is instructive to see precisely what is involved in carrying it out. We have

$$\frac{d}{dt}[q(t)e^{t/CR}] \equiv e^{t/CR}\frac{dq}{dt} + \frac{1}{CR}e^{t/CR}q(t) = \frac{v(t)}{R}e^{t/CR}$$

so that

$$q(t)e^{t/CR} - q(t_0)e^{t_0/CR} = \int_{t_0}^{t} \frac{v(\tau)}{R}e^{\tau/CR}d\tau. \tag{4.33}$$

If we make the very reasonable assumption that the charge q_0 remains bounded as t_0 tends to $-\infty$, then we must have

$$\lim_{t_0 \to -\infty}[q(t_0)e^{t_0/CR}] = 0.$$

Hence, allowing t_0 to tend to $-\infty$ in (4.33) we get

$$q(t) = \frac{1}{R}\int_{-\infty}^{t} v(\tau)e^{-(t-\tau)/CR}d\tau \tag{4.34}$$

which we can write in the form

$$q(t) = \int_{-\infty}^{t} v(\tau)h(t-\tau)d\tau \equiv v(t) \star h(t) \tag{4.35}$$

where

$$h(t) = \frac{u(t)}{R}e^{-t/CR}.$$

This expresses q explicitly in terms of v and the relationship can once again be written in operator form:

$$q = Tv. \tag{4.36}$$

The operator T is linear and time-invariant and admits the convolution integral representation of (4.34). The function h is the impulse response function and represents the charge on the capacitor due to an impulsive (delta function) voltage excitation at time $t = 0$. Since $h(t)$ vanishes for all $t < 0$ the system is a causal one, as indeed the physical background of the situation would lead us to expect.

4.5.5 RC circuit transfer function. In practice we are most often interested in examining the behaviour of the system from some given instant onwards, say for all $t > 0$. Taking $t_0 = 0$ in (4.33) we would get

$$q(t) = \frac{1}{R} \int_0^t v(\tau) e^{-(t-\tau)/CR} d\tau + q_0 e^{-t/CR} \qquad (4.37)$$

where we write q_0 for $q(0)$, the charge on the capacitor at time $t = 0$. If v is taken to be a 'suddenly applied' excitation voltage, so that $v(t) = 0$ for all $t < 0$, then q_0 is an arbitrarily specified quantity which represents the effect of any energy stored in the system at time $t = 0$. Note that unless this initial charge is zero the relationship between q and v expressed by (4.37) is not linear. Thus suppose that the input voltage is a unit step, say $v_1(t) = u(t)$. Then for $t > 0$ we get

$$q_1(t) = q_0 e^{-t/CR} + \frac{e^{-t/CR}}{R} \int_0^t e^{\tau/CR} d\tau$$

$$= q_0 e^{-t/CR} + C[1 - e^{-t/CR}].$$

On the other hand, if $v_2(t) = 2u(t)$ then

$$q_2(t) = q_0 e^{-t/CR} + 2C[1 - e^{-t/CR}] \neq 2q_1(t) \quad \text{unless} \quad q_0 = 0.$$

The classical Laplace Transform is peculiarly well suited to problems in which initial conditions must be taken into account. In the next chapter we shall give a brief review of the classical, one-sided, Laplace Transform (hereinafter referred to simply as the Laplace Transform) together with its natural extension to the generalised functions introduced in what has gone before. In the particular example discussed here note that we can obtain the Laplace Transform of the function h or, equivalently, the Transfer Function of the (causal) system of which h is the impulse response, by computing directly the system response to the input $v(t) = e^{st}$. Thus from (4.34) we get

$$q(t) = \frac{1}{R} \int_{-\infty}^t e^{s\tau} e^{-(t-\tau)/CR} d\tau = \frac{C}{1 + sCR} e^{st}$$

so that the transfer function is given by

$$H(s) = \frac{C}{1 + sCR} \equiv \mathcal{L}[h(t)].$$

Further reading.

Many texts on linear systems, network theory, signal processing etc. are currently available, offering basic information on the role of the delta function as unit impulse and the relation with integral transforms, transfer function and so on. Among the more recent is **Signal Processing in Electronic Communication** by **Michael Chapman, David Goodall** and **Nigel Steel**, published in the Horwood Series on Engineering Science in 1997. This gives an up-to-date account of linear systems theory which enlarges considerably on the introductory material presented in this chapter.

In addition, however, it is worth drawing attention to the much earlier book **Mathematics of Dynamical Systems** by **H.H.Rosenbrock** and **C.Storey**. This has already been recommended in Chapter 2 above for its unusually detailed and carefully prepared alternative theory of the delta function as a Stieltjes integral. We recommend it again here for its very individual, rigorous and elegant treatment of the fundamental theory of linear, time-invariant systems. Once again the reader should be warned of another variation in the notation for the Heaviside unit step. Like van der Pol and Bremmer, the authors settle for the symbol U, but this time with $U(t)$ defined to be zero at $t = 0$. More importantly, their treatment of the delta function introduces a fundamental difference in significance between expressions $\delta(t - a)$ and $\delta(a - t)$. These would seem to need the rather unhappy interpretation of an "impulse at $a+$" (that is, "just before a"), and an "impulse at $a-$" (that is, "just after a"). Nevertheless, despite the resulting conceptual obscurity, the account devised by Rosenbrock and Storey amply repays serious study.

Chapter 5

The Laplace Transform

5.1 THE CLASSICAL LAPLACE TRANSFORM

5.1.1 Laplace transform integral. Let f be a function of the real variable t which is defined for all $t \in \mathbb{R}$ and which is either continuous or at least sectionally continuous for all $t \geq 0$. The classical Laplace Transform of f is the function $F_0(s) \equiv \mathcal{L}[f(t)](s)$ defined by the formula

$$F_0(s) \equiv \mathcal{L}[f(t)](s) = \int_0^{+\infty} f(t)e^{-st}dt. \tag{5.1}$$

We use the notation $F_0(s)$ for this one-sided Laplace transform of f, and reserve $F(s)$ for the two-sided Laplace Transform defined by some authorities as,

$$F_0(s) \equiv \mathcal{L}[f(t)](s) = \int_{-\infty}^{+\infty} f(t)e^{-st}dt.$$

This is consistent with our later use of $F(i\omega)$ for the Fourier Transform. Recall, in passing, that the classical one-sided transform of a function $f(t)$ could always be regarded as the two-sided transform of the function $f_0(t) \equiv u(t)f(t)$,

$$F_0(s) = \int_{-\infty}^{+\infty} u(t)f(t)e^{-st}dt.$$

The definition of $F_0(s)$ clearly makes sense only for those values of s for which the infinite integral (5.1)is convergent. For many applications it is enough to regard s as a real parameter, but in general it should be taken as complex, say $s = \sigma + i\omega$. Thus F_0 is really a function of a complex variable defined over a certain region of the complex plane; the region of definition comprises just those values of s for which the infinite integral exists.

5.1.2 Existence of the classical transform. For any specific value of s the infinite integral defining $F_0(s)$ will converge if $|f(t)|$ does not increase 'too rapidly' as t goes to infinity. A sufficient condition for the existence of $F_0(s)$ can be stated as follows:

Existence Theorem.

If real numbers $M > 0$ and γ exist such that

$$|f(t)| \leq Me^{\gamma t} \quad \text{for all } t \geq T,$$

then $F_0(s)$ exists for all s such that $\text{Re}(s) > \gamma$. (In such a case f is said to be a **function of exponential order** as $t \to \infty$.)

Proof.

$$F_0(s) = \int_0^{+\infty} e^{-st}f(t)dt = \int_0^{T} e^{-st}f(t)dt + \int_T^{+\infty} e^{-st}f(t)dt$$

and

$$\left| \int_T^{+\infty} e^{-st}f(t)dt \right| \leq \int_T^{+\infty} |e^{-st}f(t)|dt \leq \int_T^{+\infty} e^{-\sigma t}|f(t)|dt$$

$$\leq \int_T^{+\infty} Me^{-(\sigma-\gamma)t}dt \leq \int_0^{+\infty} Me^{-(\sigma-\gamma)t}dt.$$

Provided that $\sigma \equiv \text{Re}(s) > \gamma$, the last integral has the finite value $M/(\sigma - \gamma)$. Hence $F_0(s)$ certainly exists for any s such that $\text{Re}(s) > \gamma$, and the defining integral (5.1) converges absolutely for all such values of the parameter s.

Unless otherwise stated we shall assume throughout that every function $f(t)$ which has a well-defined Laplace Transform $F_0(s)$ (over some region of the complex plane) is a function of exponential order as $t \to \infty$. In particular this means that we always have

$$\lim_{t \to \infty} e^{-st}f(t) = 0 \tag{5.2}$$

for every value of s at which $F_0(s)$ is defined.

5.1.3 Fundamental properties of the classical transform. Elementary applications of the Laplace Transform depend essentially on three basic properties:

(i) **Linearity.** If the Laplace Transforms of f and g are $F_0(s)$ and $G_0(s)$ respectively, and if a_1 and a_2 are any (real) constants, then the Laplace Transform of the function h defined by

$$h(t) = a_1 f(t) + a_2 g(t)$$

is

$$H_0(s) = a_1 F_0(s) + a_2 G_0(s).$$

The proof is trivial.

(ii) **Transform of a Derivative.** If f is differentiable (and therefore continuous) for $t \geq 0$, then

$$\mathcal{L}[f'(t)](s) = s F_0(s) - f(0). \tag{5.3}$$

Proof. Using integration by parts we have

$$\mathcal{L}[f'(t)](s) = \int_0^{+\infty} e^{-st} f'(t) dt$$

$$= \left[e^{-st} f(t) \right]_0^{+\infty} + \int_0^{+\infty} s e^{-st} f(t) dt$$

$$= -f(0) + s \int_0^{+\infty} e^{-st} f(t) dt$$

since $\lim_{t \to \infty} e^{-st} f(t) = 0$.

Corollary. If f is n-times differentiable for $t \geq 0$ then

$$\mathcal{L}[f^{(n)}(t)] = s^n F_0(s) - s^{n-1} f(0) - s^{n-2} f'(0) - \ldots - f^{n-1}(0).$$

(iii) **The Convolution Theorem.** Let f and g have Laplace Transforms $F_0(s)$ and $G_0(s)$ respectively, and define h as follows:

$$h(t) = \int_0^t f(\tau) g(t - \tau) d\tau, \quad t \geq 0.$$

Then,

$$\mathcal{L}[h(t)] = F_0(s)G_0(s). \tag{5.4}$$

Proof. The Laplace Transform of h is given by

$$H_0(s) = \int_0^{+\infty} e^{-st} \left[\int_0^t f(\tau)g(t-\tau)d\tau \right] dt.$$

Now,

$$\int_0^t f(\tau)g(t-\tau)d\tau = \int_0^{+\infty} f(\tau)g(t-\tau)u(t-\tau)d\tau$$

because

$$u(t-\tau) = 1 \qquad \text{for all } \tau \text{ such that } \tau < t$$

and

$$u(t-\tau) = 0 \qquad \text{for all } \tau \text{ such that } \tau > t.$$

Hence

$$H_0(s) = \int_0^{+\infty} e^{-st} \left[\int_0^{+\infty} f(\tau)g(t-\tau)u(t-\tau)d\tau \right] dt.$$

Again,

$$\int_0^{+\infty} g(t-\tau)u(t-\tau)e^{-st}dt = \int_\tau^{+\infty} g(t-\tau)e^{-st}dt$$

because

$$u(t-\tau) = 1 \qquad \text{for all } t \text{ such that } t > \tau,$$

and

$$u(t-\tau) = 0 \qquad \text{for all } t \text{ such that } t < \tau.$$

Thus,

$$H_0(s) = \int_0^{+\infty} f(\tau) \left[\int_\tau^{+\infty} g(t-\tau)e^{-st}dt \right] d\tau$$

and so, putting $T = t - \tau$, we get

$$H_0(s) = \int_0^{+\infty} f(\tau) \left[\int_0^{+\infty} g(T)e^{-s(T+\tau)}dT \right] d\tau$$

since $T = 0$ when $t = \tau$. That is,

$$H_0(s) = \int_0^{+\infty} f(\tau)e^{-s\tau}d\tau \int_0^{+\infty} g(T)e^{-sT}dT = F_0(s)G_0(s).$$

Remarks. (i) The change in the order of integration in the proof given above is justified by the absolute convergence of the integrals concerned.

(ii) Recall that h, as defined here, is the convolution of the functions $u(t)f(t)$ and $u(t)g(t)$. If f and g happen to be functions which vanish identically for all negative values of t then the above result can be expressed in the form,

The Laplace Transform of the convolution of f and g is the product of the individual Laplace transforms.

5.1.4 Application to differential equations. The most immediate application of the above properties is in the solution of ordinary differential equations with constant coefficients. Consider the case of the general second-order equation

$$a\frac{d^2y}{dt^2} + 2b\frac{dy}{dt} + cy = f(t) \tag{5.5}$$

where $y(0) = \alpha$ and $y'(0) = \beta$. If $\mathcal{L}[y(t)](s) = Y_0(s)$ then

$$\mathcal{L}\left[\frac{dy}{dt}\right](s) = sY_0(s) - \alpha, \quad \text{and} \quad \mathcal{L}\left[\frac{d^2y}{dt^2}\right](s) = s^2Y_0(s) - \alpha s - \beta.$$

Taking Laplace Transforms of both sides of (5.5) therefore gives

$$a[s^2Y_0(s) - \alpha s - \beta] + 2b[sY_0(s) - \alpha] + cY_0(s) = F_0(s).$$

That is,

$$Y_0(s) = \frac{F_0(s)}{as^2 + 2bs + c} + \frac{a\alpha s + (a\beta + 2b\alpha)}{as^2 + 2bs + c}. \tag{5.6}$$

$Y_0(s)$ is thus given explicitly as a function of s, and what remains is an **inversion problem**: that is to say we need to determine a function $y(t)$ whose Laplace Transform is $Y_0(s)$. The question of uniqueness which naturally arises is not, in practice, a serious problem. In brief, if y_1 and y_2 are any two functions which have the same Laplace Transform $Y_0(s)$, then they can differ in value only on a set of points which is (in a sense which can be made precise) negligibly small. In fact we have the following situation

$$\text{if} \quad \mathcal{L}[y_1(t)] = \mathcal{L}[y_2(t)] \quad \text{then} \quad \int_0^{+\infty} |y_1(t) - y_2(t)|\,dt = 0,$$

and it follows in particular that $y_1(t) = y_2(t)$ at least at all points of continuity. With this proviso in mind, we admit the slight abuse of notation involved and write

$$y(t) \equiv \mathcal{L}^{-1}[Y_0(s)] = \mathcal{L}^{-1}\left[\frac{F_0(s)}{as^2 + 2bs + c}\right]$$

$$+\mathcal{L}^{-1}\left[\frac{a\alpha s + (a\beta + 2b\alpha)}{as^2 + 2bs + c}\right] \qquad (5.7)$$

where y is defined for all $t > 0$. A more serious problem from the practical point of view is that of implementing the required inversion; that is, of devising effective procedures which allow us to recover a function $f(t)$ given its Laplace Transform $F_0(s)$. In a large number of commonly occurring cases this can be done by expressing $F_0(s)$ as a combination of standard functions of s whose inverse transforms are known (see Sec. 5.4 below).

5.1.5 Particular integral and complementary function. Note that with zero initial conditions, $(y(0) = 0 = y'(0))$, the differential equation (5.5) can be regarded as representing a linear time-invariant system which transforms a given input signal f into a corresponding output y. This output function y is the **particular integral** associated with f and, using the Convolution Theorem, it can be expressed in terms of the appropriate impulse response function characterising the system:

$$y(t) = \int_0^t f(\tau)h(t - \tau)d\tau = \hat{L}^{-1}[F_0(s)H_0(s)]$$

where

$$H_0(s) = \int_0^{+\infty} e^{-st}h(t)dt = \frac{1}{as^2 + 2bs + c}.$$

Non-zero initial conditions correspond to the presence of stored energy in the system at time $t = 0$. The response of the system to this stored energy is independent of the particular input f and is given by the **complementary function**. The complete solution (valid for all $t > 0$) of the equation (5.5) can be written in the form

$$y(t) = \mathcal{L}^{-1}[F_0(s)H_0(s)] + \mathcal{L}^{-1}[\{a\alpha s + (a\beta + 2b\alpha)\}H_0(s)]. \qquad (5.8)$$

In applying the classical Laplace Transform technique to (5.5) we are tacitly assuming that the system which it is being taken to represent is **unforced** for $t < 0$: that is, that the response which we compute from (5.8) is actually the response to the excitation $u(t)f(t)$. This is sometimes expressed by saying that the input is **suddenly applied** at $t = 0$.

Exercises I.

1. Find (from first principles, or otherwise) the Laplace Transforms of each of the following functions:

(a) $u(t)$; (b) $u(t) - u(t-1)$; (c) e^{at} ; (d) t ;

(e) $\cosh(bt)$; (f) $\sin(\omega t)$; (g) $\cos(\omega t)$.

2. Use Laplace Transforms to obtain, for $t > 0$, the solution of the linear differential equation

$$\frac{d^2y}{dt^2} + y = t$$

which satisfies the conditions $y(0) = 1$, $y'(0) = -2$.

3. Use the Convolution Theorem for the Laplace transform to solve the integral equation

$$y(t) = \cos(t) + 2\sin(t) + \int_0^t y(\tau)\sin(t-\tau)d\tau$$

for $t > 0$.

4. Solve the simultaneous linear differential equations,

$$\frac{dx}{dt} = 2x - 3y$$

$$\frac{dy}{dt} = y - 2x$$

for $t > 0$, given that $x(0) = 8$, and $y(0) = 3$.

5. A causal, time-invariant linear system has impulse response

$$g(t) = \frac{u(t)}{\sqrt{\pi t}}.$$

Find its frequency response function. (Hint. See Exercises II, No. 4, of Chapter 4.)

5.2   LAPLACE   TRANSFORMS   OF
DELTA FUNCTIONS

5.2.1 Transforms of delta functions. In Chapter 4 the delta function and its derivatives were seen to play essential roles in the analysis of linear systems, appearing both as signals and as (generalised) impulse response functions for certain special systems. It is therefore necessary to extend the definition of the Laplace Transform to apply to such generalised functions. In general we should expect to be able to define the transform of any $f(t)$ of the form

$$f(t) = f_r(t) + f_s(t)$$

where the **regular part**, $f_r(t)$, denotes a function (in the proper sense) for which the classical Laplace transform is well-defined by the Laplace integral (5.1), and the **singular part**, $f_s(t)$, is some linear combination of delta functions and/or derivatives of delta functions.

If $a \neq 0$ then there is no especial difficulty in assigning a meaning to the Laplace transform of the delta function $\delta(t - a)$ located at the point $t = a$. A direct application of the sampling property is permissible because $e^{-st}u(t)$ is well-defined and continuous at $t = a$, and we have at once that, when $a > 0$,

$$\mathcal{L}[\delta(t-a)] = \int_0^{+\infty} e^{-st}\delta(t-a)dt \equiv \int_{-\infty}^{+\infty} e^{-st}u(t)\delta(t-a)dt = e^{-sa}. \quad (5.9)$$

The result when $a < 0$ is obviously zero.

Again, for the derivative $\delta'(t - a)$, where $a > 0$, we would get

$$\mathcal{L}[\delta'(t - a)] = \int_{-\infty}^{+\infty} e^{-st}u(t)\delta'(t)dt = -\left[\frac{d}{dt}(e^{-st})\right]_{t=a} = se^{-sa},$$

and similarly for higher derivatives of $\delta(t - a)$, the general result being,

$$\mathcal{L}[\delta^{(n)}(t - a)] = s^n e^{-sa}. \quad (5.10)$$

However, as the reader will be well aware by this time, there is a serious complication if $a = 0$ and we then try to attach a meaning to the Laplace transform of a delta function, $\delta(t)$, located at the origin itself. The existing rules for the sampling property of $\delta(t)$ fail in the case of the Laplace integral

$$\int_0^{+\infty} e^{-st}\delta(t)dt \equiv \int_{-\infty}^{+\infty} e^{-st}u(t)\delta(t)dt,$$

since there is a discontinuity of the function $e^{-st}u(t)$ at the point $t = 0$. As before we are obliged to accept that we can only say that the formal product $u(t)\delta(t)$ is equivalent to $k\delta(t)$, for some arbitrary constant k, (see the dicussion in Sec. 3.6.2). This would allow us to write $\mathcal{L}[\delta(t)] = k$, but would still leave the Laplace transform of $\delta(t)$ not uniquely defined.

This difficulty has already been referred to in Sec. 4.5, where the role of the delta function as a (generalised) impulse response function suggested that we should set $\mathcal{L}[\delta(t)] = 1$ for all s. This is consistent with the result obtained in (5.9) above for the Laplace transform of $\delta(t - a)$ if we allow a to tend to zero, and it is, in fact, the definition most usually adopted. Accordingly we shall settle for the following formal definition:

Definition of $\mathcal{L}[\delta(t)]$:

$$\mathcal{L}[\delta(t)] = \lim_{a\downarrow 0}\mathcal{L}[\delta(t - a)] = 1$$

and similarly,

$$\mathcal{L}\delta^{(n)}(t) = \lim_{a\downarrow 0}\mathcal{L}[\delta^{(n)}(t - a)] = s^n.$$

Nevertheless it should be realised that other values for $\mathcal{L}[\delta(t)]$ could be chosen, and indeed have been seriously suggested. In particular, as shown in Sec. 3.2.3 and in Sec. 3.6.2 earlier, there are persuasive arguments for interpreting the formal product $u(t)\delta(t)$ as equivalent to $\frac{1}{2}\delta(t)$. This would produce the result

$$\int_0^{+\infty} \delta(t)dt \equiv \int_{-\infty}^{+\infty} u(t)\delta(t)dt = 1/2,$$

which has a certain popularity. But this choice would also imply that

$$\int_0^{+\infty} \delta(t)e^{-st}dt \equiv \int_{-\infty}^{+\infty} u(t)\delta(t)e^{-st}dt = 1/2,$$

and this would have significant effects on the development of a transform algebra. In order to discuss the effect of alternative choices for the definition of the Laplace transform of $\delta(t)$ we shall, where necessary, temporarily adopt a neutral symbolism in what follows, and write

$$\mathcal{L}[\delta(t)] \equiv \Delta(s). \tag{5.11}$$

5.2.2 Transform of a function discontinuous at a point $a > 0$.
We have previously established a derivative rule (5.3) for the Laplace transform of a function which is differentiable in the classical sense on $[0, +\infty)$:

$$\mathcal{L}[f'(t)] = s \int_0^{+\infty} f(t)dt - f(0)$$
$$\equiv sF_0(s) - f(0).$$

Once we admit functions with a singular part, the difference between differentiation in the classical sense and in what we have termed the generalised, or operational, sense must be taken into account. In particular we need to examine first what happens in the case of a function with simple discontinuities. Accordingly, let f be a function which is continuously differentiable everywhere in the classical sense except at a point $a > 0$ at which it has a jump discontinuity.
As in Sec. 3.3 we can write f as

$$f(t) = \phi_1(t)u(a - t) + \phi_2(t)u(t - a), \tag{5.12}$$

where $a > 0$, and ϕ_1 and ϕ_2 are everywhere continuously differentiable functions. The generalised derivative $Df(t)$ for $f(t)$ contains a delta function located at the point $t = a$ whose strength is given by the jump, or saltus, of $f(t)$ at that point:

$$Df(t) = \phi_1'(t)u(a - t) + \phi_2'(t)u(t - a) + [\phi_2(a) - \phi_1(a)]\delta(t - a)$$
$$\equiv f'(t) + [f(a+) - f(a-)]\delta(t - a). \tag{5.13}$$

If we wish to find the Laplace transform of $Df(t)$ then, as the discussion in the preceding section makes clear, the term in (5.13) involving the delta function at $t = a$ presents no problem. We can write

$$\int_0^{+\infty} e^{-st}[Df(t)]dt = \int_0^{+\infty} e^{-st}f'(t)dt + [f(a+) - f(a-)]e^{-sa}. \tag{5.14}$$

However, we then need to compute the Laplace transform of the classical derivative $f'(t)$ of $f(t)$. We can no longer do this by a simple appeal to the derivative rule given above since this specifically assumes that $f(t)$ is classically differentiable and therefore continuous. Instead we must carry out a direct calculation of the transform of $f'(t)$, as follows:

$$f'(t) = \phi_1'(t)u(a-t) + \phi_2'(t)u(t-a) \quad \text{(for all } t \neq a),$$

so that

$$\mathcal{L}[f'(t)] = \int_0^{+\infty} e^{-st} f'(t)dt = \int_0^a \phi_1'(t)e^{-st}dt + \int_a^{+\infty} \phi_2'(t)e^{-st}dt$$

$$= \left[e^{-st}\phi_1(t)\right]_0^a + s\int_0^a \phi_1(t)e^{-st}dt$$

$$+ \left[e^{-st}\phi_2(t)\right]_a^{+\infty} + s\int_a^{+\infty} \phi_2(t)e^{-st}dt$$

$$= s\left[\int_0^a \phi_1(t)e^{-st}dt + \int_a^{+\infty} \phi_2(t)e^{-st}dt\right]$$

$$-e^{-sa}[\phi_2(a) - \phi_1(a)] - \phi_1(0)$$

That is,

$$\int_0^{+\infty} e^{-st} f'(t)dt == sF_0(s) - f(0) - e^{-sa}[f(a+) - f(a-)], \quad (5.15)$$

which is significally different from the standard derivative rule (5.3). But, substituting for $\int_0^{+\infty} e^{-st} f'(t)dt$ in equation (5.14), we have,

$$\int_0^{+\infty} e^{-st}[Df(t)]dt = sF_0(s) - f(0), \quad (5.16)$$

and this result for the Laplace transform of a generalised derivative does have the same form as that of the standard derivative rule (5.3).

5.2.3 Transform of a function discontinuous at the origin. If we allow the point of discontinuity, $a > 0$, to tend to zero then the derivative rule for the transform of the classical derivative of $f(t)$ is easily computed since then we have

$$f(t) = \phi_1(t)u(-t) + \phi_2(t)u(t)$$

and so,

$$\mathcal{L}[f'(t)](s) = \int_0^{+\infty} \phi_2'(t)e^{-st}dt = s\int_0^{+\infty} \phi_2(t)e^{-st}dt - \phi_2(0).$$

That is,

$$\mathcal{L}[f'(t)](s) = sF_0(s) - f(0+). \qquad (5.17)$$

But then

$$Df(t) = \phi_1'(t)u(-t) + \phi_2'(t)u(t) + [\phi_2(0) - \phi_1(0)]\delta(t).$$

so that for the transform of the generalised derivative we get

$$\mathcal{L}[Df(t)] = \mathcal{L}[\phi_2'(t)] + [\phi_2(0) - \phi_1(0)]\mathcal{L}[\delta(t)]$$

$$= s\mathcal{L}[\phi_2(t)] - \phi_2(0) + [\phi_2(0) - \phi_1(0)]\Delta(s)$$

$$\equiv sF_0(s) - f(0+) + [f(0+) - f(0-)]\Delta(s). \qquad (5.18)$$

and at this stage the choice of a value for the Laplace transform, $\Delta(s)$, of $\delta(t)$ becomes crucial.

The good news is that whatever value we choose for $\Delta(s)$, equation (5.18) is bound to be consistent with the behaviour of δ as the derivative of the unit step function $u(t)$. For, since

$$\mathcal{L}[u(t)] = \int_0^{+\infty} e^{-st}dt = 1/s,$$

we have

$$\mathcal{L}[Du(t)] = [s(1/s) - u(0+)] + \Delta(s)[u(0+) - u(0-)]$$

$$= (1 - 1) + \Delta(s)(1 - 0) = \Delta(s).$$

On the other hand care must be taken to ensure that, for an arbitrary function $f(t)$, the correct form of the derivative rule, (5.19), is used once a specific definition of $\Delta(s)$ has been decided on. Thus, for the definition $\Delta(s) = 1$ that we have actually agreed upon, we should have

$$\mathcal{L}[Df(t)] = sF_0(s) - f(0-) = sF_0(s) - f(0-) \qquad (5.19).$$

In particular, if $f(t)$ is a function which vanishes for all negative values of t then this becomes

$$\mathcal{L}[Df(t)] = sF_0(s),$$

and similarly,

$$\mathcal{L}[D^2f(t)] = s^2F_0(s) - sf(0-) - f'(0-) = s^2F_0(s).$$

Had the choice been for $\Delta(s) = 1/2$, then the result should have been

$$\mathcal{L}[Df(t)] = sF_0(s) - \frac{1}{2}[f(0+) - f(0-)],$$

while for the somewhat bizarre (but technically possible) definition $\Delta(s) = 0$, we would have to use

$$\mathcal{L}[Df(t)] = sF_0(s) - f(0+).$$

5.2.4 Initial conditions and the $0+/0-$ controversy. We have used throughout a straightforward definition of the classical Laplace transform as a simple integral, (5.1), with limits 0 and $+\infty$, and supplemented this with ad hoc formal definitions of transforms of the delta function and its derivatives. This has shown that there are distinct variants of the derivative rule (5.3) which depend on whether or not $f(t)$ has any simple discontinuities, and whether the classical or generalised derivative is being considered. In order to interpret these results and to obtain a consistent use of information about initial conditions, many authorities have chosen to modify the basic definition of the Laplace integral itself so as to incorporate the effects of discontinuities and allow for the presence of delta functions. To deal with possible discontinuities at the origin, two alternative definitions of Laplace transform have therefore become established in the literature. In place of the Laplace integral (5.1) we have either,

$$\mathcal{L}_+[f(t)] = \int_{0+}^{+\infty} f(t)e^{-st}dt \equiv \lim_{\epsilon \downarrow 0} \int_{\epsilon}^{+\infty} f(t)e^{-st}dt \qquad (5.20a)$$

or

$$\mathcal{L}_-[f(t)] = \int_{0-}^{+\infty} f(t)e^{-st}dt \equiv \lim_{\epsilon \downarrow 0} \int_{-\epsilon}^{+\infty} f(t)e^{-st}dt \qquad (5.20b)$$

A good deal of confusion has been generated by these two conflicting definitions, and there has been much controversial discussion of their apparent significance in applications of the Laplace transform.

It is generally appreciated that definition (5.20a) necessarily implies the use of the derivative rule (5.17) which, as we have shown, is associated with the use of the classical derivative $f'(t)$. It appears to be less generally appreciated that this carries with it the implication that $\Delta(s)$ must be taken as zero! To recover from this in applications which, on physical grounds, seem to demand a viable delta function to be present at $t = 0$, a popular, but rather desperate, solution is to assert that $\delta(t)$ is to be located "just to the right"of the origin. But no such interpretation of the conventional symbol 0+ makes sense, at least within the standard real number system.

By contrast the use of (5.20b), coupled with the derivative rule (5.19), which is properly associated with use of the generalised derivative, $Df(t)$, produces a transform calculus in which $\Delta(s) = 1$ and which is wholly consistent with that developed here. The only point at issue is whether the introduction of an extended integration convention such as is given in (5.20b) is actually necessary, or desirable. It could be argued that the formalism of (5.20b) does give students a convincing picture of the inclusion of $\delta(t)$ within the range of integration, although it encourages the myth of a location "just to the left"of the origin. The alternative treatment, as developed in this text, leaves intact the usual, standard symbolism of the integral calculus at the expense of adopting a distinctive notation for the very significant generalisation of differentiation which is at the heart of the matter.

We shall return briefly to the possible significance of the $0+/0-$ symbolism in the final chapter (on Nonstandard Analysis) of this book. For the moment we refer the reader to the very lucid and comprehensive analysis of the whole subject of the $0 + /0-$ controversy in the paper by Lundberg, Miller and Trumper quoted in the suggested recommendations for Further Reading at the end of this chapter.

5.2.5 Transform of a periodic function. An additional example of the value of the definition of the Laplace transform of the delta function as 1 is to be seen in the derivation of the Laplace transform of a **periodic** function:

Let f be a function which vanishes identically outside the finite interval $(0, T)$. The **periodic extension** of f, of period T, is defined to be the function obtained by summing the translates, $f(t - kT)$, for $k = 0, \pm 1, \pm 2, \ldots$, (see Fig. 5.1),

$$f_T(t) = \sum_{k=-\infty}^{+\infty} f(t - kT). \tag{5.21}$$

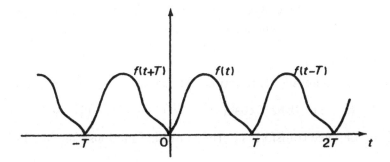

Figure 5.1:

The Laplace transform of the translate $f(t - kT)$ is $F_0(s)e^{-kTs}$ and it is easy to see that

$$\mathcal{L}[f_T(t)] = \sum_{k=0}^{+\infty} F_0(s)e^{-kTs} = \frac{F_0(s)}{1 - e^{-sT}}, \qquad (5.22)$$

the summation being valid provided that

$$|e^{-sT}| = |e^{-(\sigma + i\omega)t}| = e^{-\sigma T} < 1$$

that is, for all s such that $Re(s) > 0$.

Now if f is continuous on $(0, T)$ then we can write f_T as a convolution:

$$f_T(t) = \sum_{k=-\infty}^{+\infty} [f(t) \star \delta(t - kT)] = f(t) \star \sum_{k=-\infty}^{+\infty} \delta(t - kT). \qquad (5.23)$$

Further, using the agreed definition of $\Delta(s)$ as unity, we obtain

$$\mathcal{L}\left[\sum_{k=-\infty}^{+\infty} \delta(t - kT)\right] = \mathcal{L}\left[\sum_{k=0}^{+\infty} \delta(t - kT)\right]$$

$$= 1 + e^{-sT} + e^{-2sT} + e^{-3sT} + \ldots = \frac{1}{1 - e^{-sT}}, \qquad (5.24)$$

the summation again being valid provided that $Re(s) > 0$. Hence, (5.22) and (5.23) actually show that

$$\mathcal{L}\left[f(t) \star \sum_{-\infty}^{+\infty} \delta(t - kT)\right] = F_0(s) \times \frac{1}{1 - e^{-sT}},$$

which confirms that the Convolution Theorem for the Laplace transform remains valid for these (generalised) factors.

Exercises II.

1. Identify the (generalised) functions whose Laplace transforms are

(a) $(s^3 + 2)/(s + 1)$; (b) $e^{-s} \cosh(s)$; (c) $(1 - s^n e^{-ns})/(1 - se^{-s})$.

2. Find the Laplace transforms of the periodic functions sketched in Fig. 5.2.

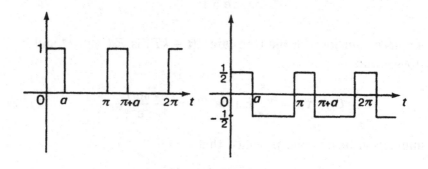

Figure 5.2:

3. Show that

$$\mathcal{L}[\delta(\cos(t)] = \frac{e^{-\pi s/2}}{1 - e^{-\pi s}}.$$

5.3 COMPUTATION OF LAPLACE TRANSFORMS

5.3.1 If f is an ordinary function whose Laplace transform exists (for some values of s) then we should be able to find that transform, in principle at least, by evaluating directly the integral which defines $F_0(s)$. It is usually simpler in practice to make use of certain appropriate properties of the Laplace integral and to derive specific transforms from them. The following results are easy to establish and are particularly useful in this respect:

(L.T.1) **The First Translation Property.** If $\mathcal{L}[f(t)](s) = F_0(s)$, and if a is any real constant, then

$$\mathcal{L}[e^{at}f(t)](s) = F_0(s-a).$$

(L.T.2) **The Second Translation Property.** If $\mathcal{L}[f(t)](s) = F_0(s)$, and if a is any positive constant, then

$$\mathcal{L}[u(t-a)f(t-a)](s) = e^{-as}F_0(s).$$

(L.T.3) **Change of Scale.** If $\mathcal{L}[f(t)](s) = F_0(s)$, and if a is any positive constant, then

$$\mathcal{L}[f(at)](s) = \frac{1}{a}F_0\left(\frac{s}{a}\right).$$

(L.T.4) **Multiplication by** t. If $\mathcal{L}[f(t)](s) = F_0(s)$, then

$$\mathcal{L}[tf(t)](s) = -\frac{d}{ds}F_0(s) \equiv -F_0'(s).$$

(L.T.5) **Transform of an Integral.** If $\mathcal{L}[f(t)](s) = F_0(s)$, and if the function g is defined by

$$g(t) = \int_0^t f(\tau)d\tau$$

then

$$\mathcal{L}[g(t)](s) = \frac{1}{s}F_0(s).$$

The first three of the above properties follow immediately on making suitable changes of variable in the Laplace integrals concerned. For (L.T.4) we have only to differentiate the Laplace transform integral of $f(t)$ with respect to the parameter s under the integral sign. Finally, in the case of (L.T.5), it is enough to note that $g'(t) = f(t)$ and that $g(0) = 0$; the result then follows from the rule for finding the Laplace transform of a derivative.

Using these properties, an elementary basic table of standard transforms can be constructed without difficulty, and is shown in Table I below. This

table can be extended by using various special techniques, in particular
by exploiting the results derived above for the Laplace transforms of delta
functions and derivatives of delta functions.

$f(t)$	$F_0(s)$	Region of (absolute) convergence
$u(t)$	$\dfrac{1}{s}$	$\mathrm{Re}(s) > 0$
t	$\dfrac{1}{s^2}$	$\mathrm{Re}(s) > 0$
$t^n\,(n > 1)$	$\dfrac{n!}{s^{n+1}}$	$\mathrm{Re}(s) > 0$
e^{at}	$\dfrac{1}{s-a}$	$\mathrm{Re}(s) > a$
e^{-at}	$\dfrac{1}{s+a}$	$\mathrm{Re}(s) > -a$
$\sinh at$	$\dfrac{a}{s^2-a^2}$	$\mathrm{Re}(s) > |a|$
$\cosh at$	$\dfrac{s}{s^2-a^2}$	$\mathrm{Re}(s) > |a|$
$\sin at$	$\dfrac{a}{s^2+a^2}$	$\mathrm{Re}(s) > 0$
$\cos at$	$\dfrac{s}{s^2+a^2}$	$\mathrm{Re}(s) > 0$

Figure 5.3: Table of Laplace transforms

5.3.2 Example 1. Laplace transform of a triangular waveform.

The waveform shown in Fig 5.4 is formed by periodically extending the
basic triangular function defined on the interval $[0, 2]$. Once the Laplace
transform of this function $f(t)$ has been determined we can derive the
Laplace transform of the whole wave by applying the formula (5.24) for
the transform of the periodic extension $\sum\limits_{k=-\infty}^{+\infty} f(t - 2k)$. To find $\mathcal{L}[f(t)]$
note first that, as shown in Fig 5.5, $f(t)$ can be decomposed into a linear

combination of ramp functions, so that we have

$$f(t) = tu(t) - 2(t-1)u(t-1) + (t-2)u(t-2),$$

and the transforms of these components of $f(t)$ are easily found.

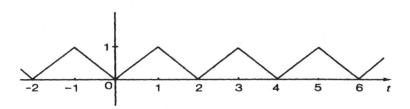

Figure 5.4:

In fact straightforward application of the second translation property (L.T.2) immediately gives

$$F_0(s) = \frac{1}{s^2} - \frac{2}{s^2}e^{-s} + \frac{e^{-2s}}{s^2} = \left[\frac{1 - e^{-s}}{s}\right]^2 = \frac{4}{s^2}e^{-s}\sinh^2\left(\frac{s}{2}\right).$$

Hence, applying (5.24),

$$\mathcal{L}[f_T(t)](s) = \left[\frac{4}{s^2}e^{-s}\sinh^2\left(\frac{s}{2}\right)\right]\left[\frac{1}{1 - e^{-2s}}\right]$$

$$= \frac{2\sinh^2(s/2)}{s^2\sinh(s)} = \frac{\tanh(s/2)}{s^2}.$$

However, brief as this calculation is, it is worthwhile examining an alternative attack which relies on the use of delta functions and of the differentiation formula (5.19). We have only to differentiate f twice to eliminate all ordinary functions and leave nothing but delta functions:

$$\frac{d}{dt}f(t) = u(t) - 2u(t-1) + u(t-2),$$

$$\frac{d^2}{dt^2}f(t) = \delta(t) - 2\delta(t-1) + \delta(t-2).$$

Hence,
$$\mathcal{L}\left[\frac{d^2}{dt^2}f(t)\right] = 1 - 2e^{-s} + e^{-2s}.$$

But, $f(t) = 0$ for all $t < 0$ and so, by (5.19),
$$\mathcal{L}\left[\frac{d^2}{dt^2}f(t)\right] = s^2 F_0(s).$$

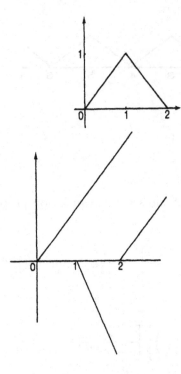

Figure 5.5:

That is,
$$F_0(s) = \frac{1}{s^2}\mathcal{L}\left[\frac{d^2}{dt^2}f(t)\right] = \frac{1 - 2e^{-s} + e^{-2s}}{s^2} = \frac{4e^{-s}}{s^2}\sinh^2(s/2).$$

Using equation (5.24), the triangular waveform shown in Fig. 5.5 is seen, as before, to have the following Laplace transform:
$$\frac{4e^{-s}\sinh^2(s/2)}{s^2(1 - e^{-2s})} = \frac{2\sinh^2(s/2)}{s^2\sinh(s)} = \frac{1}{s^2}\tanh(s/2).$$

5.3.3 A similar approach could be used to find the Laplace transform of the alternating triangular waveform shown in Fig. 5.6. In fact it is simpler to note that such a periodic function, exhibiting skew, or mirror, symmetry within each period, can be represented as a convolution with an alternating impulse train:

$$f(t) \star \sum_{k=-\infty}^{+\infty} (-1)^k \delta(t - kT/2) \tag{5.25}$$

where f now denotes the pulse shape in the first half-period. The transform of the impulse train is given by

$$\mathcal{L}\left[\sum_{k=-\infty}^{+\infty} (-1)^k \delta(t - kT/2)\right] = \mathcal{L}\left[\sum_{k=0}^{+\infty}(-1)^k \delta(t - kT/2)\right]$$

$$= \frac{1}{1 + e^{-sT/2}}.$$

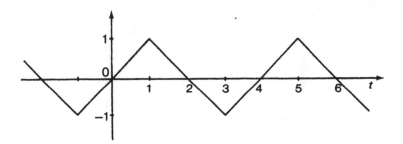

Figure 5.6:

Accordingly the Laplace transform of (5.25) is given, by the Convolution Theorem, as

$$\frac{F_0(s)}{1 + e^{-sT/2}}. \tag{5.26}$$

In particular, for the function shown in Fig. 5.6 we have the transform

$$\frac{4e^{-s} \sinh^2(s/2)}{s^2} \frac{1}{1 + e^{-2s}} = \frac{2\sinh^2(s/2)}{s^2 \cosh(s)}. \tag{5.27}$$

The technique used to compute $\mathcal{L}[f(t)]$ above can obviously be generalised. Let f be a function which vanishes for all $t < 0$ and whose graph consists of a finite number of arcs each defined by a polynomial. Then, for some value of n, the nth derivative $D^n f(t)$ will consist only of delta functions and derivatives of delta functions: this sequence of differentiations can often be done by inspection. The Laplace transform of f can then be written down by using the appropriate generalisation of (5.19),

$$\mathcal{L}\{D^n(t)\}(s) = s^n F_0(s).$$

5.3.4 Variants of the method can be used in certain situations where f does not reduce to delta function form after finitely many differentiations, as in the following example:

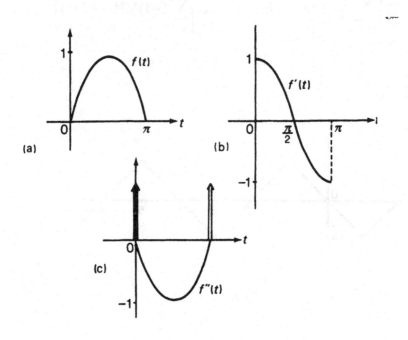

Figure 5.7:

Example 2. Let $f(t) = \sin(t)$ for $0 < t < \pi$, and $f(t) = 0$ for all other values of t. Then,

$$\frac{d}{dt} f(t) = [u(t) - u(t - \pi)] \cos(t)$$

and

$$\frac{d^2}{dt^2}f(t) = [u(t) - u(t - \pi)][-\sin(t)] + \delta(t) + \delta(t - \pi)$$

$$= \delta(t) + \delta(t - \pi) - f(t).$$

(These results can easily be obtained by inspection, as a glance at Fig. 5.7 should make clear.) Thus we can write

$$s^2 F_0(s) \equiv \mathcal{L}\left[\frac{d^2}{dt^2}f(t)\right](s) = 1 + e^{-\pi s} - F_0(s)$$

so that

$$F_0(s) = \frac{1 + e^{-\pi s}}{s^2 + 1}.$$

Applying (5.23) we obtain at once the Laplace transform of the periodic waveform shown in Fig. 5.8 ('rectified' sine wave):

$$\mathcal{L}[f_{2\pi}(t)](s) = \frac{1 + e^{-\pi s}}{s^2 + 1}\frac{1}{1 - e^{-2\pi s}} = \frac{e^{\pi s/2}}{2(s^2 + 1)\sinh(\pi s/2)}.$$

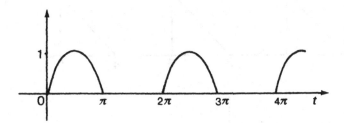

Figure 5.8:

Exercises III.

1. Find the Laplace transforms of each of the following functions:

(a) $\cosh^2(at)$; (b) $e^{-|t|}\sin^2(t)$; (c) $t^3\cosh(t)$;

(d) $t\cos(at)$; (e) $t^2\sin(at)$.

2. If $\mathcal{L}[f(t)] = F_0(s)$, show that $\mathcal{L}[f(t)/t](s) = \int_s^{+\infty} F_0(p)dp$, provided that $\lim_{t \to 0} f(t)/t$ exists. Hence find the transforms of

$$\text{(a)} \ (\sin(t)/t \ ; \quad \text{(b)} \ [\cos(at) - \cos(bt)]/t \ ; \quad \text{(c)} \ \int_0^t \frac{\sin(\tau)}{\tau} d\tau.$$

3. Find the Laplace transforms of each of the functions sketched below in Fig. 5.9(a) - (e).

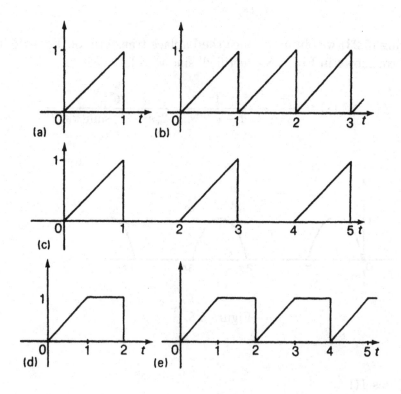

Figure 5.9:

5.4 NOTE ON INVERSION

5.4.1 There is an inversion theorem for the Laplace transform which expresses $f(t)$ explicitly in terms of $F_0(s)$. In spite of the generality of this result we shall not enlarge on it here, partly because it presupposes some acquaintance with Complex Variable theory, and partly because its use is not really necessary, or even desirable, in many elementary applications. Accordingly we confine attention to a brief account of those techniques likely to be of most value in the inversion of commonly occurring types of transforms.

Suppose first that $F_0(s) = N(s)/D(s)$, where $N(s)$ and $D(s)$ are polynomials in s with real coefficients. Without loss of generality we may assume that the degree of N is less than the degree of D; for otherwise we can divide by $D(s)$ to obtain

$$F_0(s) = P(s) + \frac{N_1(s)}{D(s)}$$

where $P(s)$ is a polynomial in s, and the degree of $N_1(s)$ is certainly lower than that of $D(s)$. Since $\mathcal{L}[\delta^{(n)}(t)] = s^n$ we can immediately identify $P(s)$ as the transform of a linear combination of delta functions and derivatives of delta functions. Similarly there is no loss of generality in assuming that $D(s)$ is normalised so that the coefficient of the highest power of s, say s^m, occurring in the denominator of $F_0(s)$ is unity. Then we may write, in general,

$$D(s) = (s - \alpha_1)^{p_1}(s - \alpha_2)^{p_2} \ldots (s - \alpha_m)^{p_m} \qquad (5.28)$$

where the zeros, α_k, may be real or complex.

5.4.2 If the zeros of $D(s)$ are simple, so that in (5.28) we have

$$p_1 = p_2 = \ldots = p_m = 1,$$

then the decomposition of $F_0(s)$ into partial fractions takes the form

$$F_0(s) \equiv \frac{N(s)}{D(s)} = \frac{A_1}{(s - \alpha_1)} + \frac{A_2}{(s - \alpha_2)} + \ldots + \frac{A_m}{(s - \alpha_m)}.$$

The numbers A_k may be determined either by equating coefficients of powers of s, or by substituting particular values of s so as to generate a system of linear algebraic equations in the A_k. Alternatively note that

$$(s - \alpha_1)\frac{N(s)}{D(s)} = A_1 + A_2\frac{s - \alpha_1}{s - \alpha_2} + \ldots + A_m\frac{s - \alpha_1}{s - \alpha_m}$$

$$(s - \alpha_2)\frac{N(s)}{D(s)} = A_1\frac{s - \alpha_2}{s - \alpha_1} + A_2 + \ldots + A_m\frac{s - \alpha_2}{s - \alpha_m}$$

and so on. It follows that

$$A_k = \lim_{s \to \alpha_k} \frac{(s - \alpha_k)N(s)}{D(s)}, \quad k = 1, 2, \ldots, m.$$

Using L'Hopital's rule for the evaluation of the limit gives the following explicit expression for A_k:

$$A_k = \lim_{s \to \alpha_k} \left[\frac{N(s) + (s - \alpha_k)N'(s)}{D'(s)}\right] = \frac{N(\alpha_k)}{D'(\alpha_k)}.$$

Since $\mathcal{L}[e^{\alpha t}](s) = 1/(s - \alpha)$, this means that we can write down the inverse transform of $F_0(s)$ in the form

$$f(t) = \frac{N(\alpha_1)}{D'(\alpha_1)}e^{\alpha_1 t} + \frac{N(\alpha_2)}{D'(\alpha_2)}e^{\alpha_2 t} + \ldots + \frac{N(\alpha_m)}{D'(\alpha_m)}e^{\alpha_m t}. \qquad (5.29)$$

The expression (5.29) applies whether the zeros of $D(s)$ turn out to be real or complex. However since $D(s)$ is a polynomial with real coefficients any complex zeros must necessarily occur in complex conjugate pairs; that is, $D(s)$ will contain, at worst, real quadratic factors of the form $s^2 + bs + c$, where $b^2 < 4c$. To rewrite (5.29) in a more compact form, free from complex exponentials, usually involves a good deal of algebraic manipulation. It is generally easier to make use directly of the existence of real quadratic factors of $D(s)$. In the partial fraction expansion such a quadratic factor gives rise to an expression like

$$\frac{As + B}{s^2 + bs + c}.$$

Now we can always write

$$s^2 + bs + c = (s + \alpha)^2 + \beta^2$$

where

$$\alpha = b/2 \text{ and } \beta^2 = \frac{1}{4}(4c - b^2).$$

Hence,

$$\frac{As + B}{s^2 + bs + c} = \frac{As + B}{(s + \alpha)^2 + \beta^2} = \frac{A(s + \alpha)}{(s + \alpha)^2 + \beta^2} + \frac{B - \alpha A}{(s + \alpha)^2 + \beta^2}.$$

Then, using the Second Translation Property (L.T.2), and recalling the standard transforms of $\sin(at)$ and $\cos(at)$, we can write down the required inverse transform as

$$Ae^{-\alpha t}\cos(\beta t) + \frac{(B - \alpha A)}{\beta}e^{-\alpha t}\sin(\beta t). \qquad (5.30)$$

5.4.3 Suppose, on the other hand, that $D(s)$ has a repeated linear factor, say $(s - \alpha)^2$. This time the partial fraction expansion will generally include the terms

$$\frac{A_{11}}{(s - \alpha)^2} + \frac{A_{12}}{(s - \alpha)}.$$

To determine A_{11} and A_{12}, note that

$$A_{11} = \lim_{s \to \alpha}\left[(s - \alpha)^2\frac{N(s)}{D(s)}\right]; \quad A_{12} = \lim_{s \to \alpha}\left[\frac{d}{ds}(s - \alpha)^2\frac{N(s)}{D(s)}\right].$$

Similarly, if $D(s)$ has a factor $(s - \alpha)^3$ then the partial fraction expansion will include the terms

$$\frac{A_{11}}{(s - \alpha)^3} + \frac{A_{12}}{(s - \alpha)^2} + \frac{A_{13}}{s - \alpha}$$

and we can determine A_{11}, A_{12} and A_{13} from

$$A_{11} = \lim_{s \to \alpha}\left[(s - \alpha)^3\frac{N(s)}{D(s)}\right]$$

$$A_{12} = \lim_{s \to \alpha}\left[\frac{d}{ds}(s - \alpha)^3\frac{N(s)}{D(s)}\right]$$

$$A_{13} = \lim_{s \to \alpha}\left[\frac{1}{2}\frac{d^2}{ds^2}(s - \alpha)^3\frac{N(s)}{D(s)}\right].$$

To invert these terms we need only appeal to the standard result

$$\mathcal{L}[t^{k-1}e^{\alpha t}](s) = \frac{(k - 1)!}{(s - \alpha)^k}.$$

In principle, similar methods could be used whenever $D(s)$ has a factor $(s - \alpha)^n$, for any positive integer n; however, for $n \geq 3$, the successive differentiations needed to determine the partial fraction coefficients as above make the work prohibitive.

5.4.4 For transforms which are not rational functions of s, more high-powered methods are usually required. However, for some types of transcendental functions simpler techniques suffice. In particular, recall that the presence of a pure exponential factor, e^{-as}, in the transform corresponds to a delay, of magnitude α, in the inverse transform. Similarly, a factor of the form $(1-e^{-sT})^{-1}$, or $(1+e^{-sT})^{-1}$, indicates that the inverse transform is a convolution with a periodic impulse train as one of the components: the resulting function may or may not be periodic. For example, it is easy to invert the following two transforms after a little elementary manipulation:

$$F_0(s) = \frac{1}{s\cosh(s)}; \quad G_0(s) = \frac{1}{s\sinh(s)}.$$

We have

$$F_0(s) = \frac{1}{s}\frac{2}{e^s + e^{-s}} = \frac{2}{se^s}\frac{1}{1 + e^{-2s}}$$

$$= \frac{2e^{-s}}{s}[1 - e^{-2s} + e^{-4s} - \ldots],$$

$$G_0(s) = \frac{1}{s}\frac{2}{e^s - e^{-s}} = \frac{2}{se^s}\frac{1}{1 - e^{-2s}}$$

$$= \frac{2e^{-s}}{s}[1 + e^{-2s} + e^{-4s} + \ldots].$$

By inspection the reader should be easily able to identify the required inverse transforms.

Further Reading.

There are a great many texts available dealing with the Laplace transform, and it is enough here just to recommend the particularly thorough modern treatment given in **Laplace Transforms and an Introduction to Distributions** by **Paul Guest**, (Ellis Horwood Series, 1991). As the title suggest this book also offers an introductory account of generalised functions which includes serious discussion of the Laplace transform of distributions. As such it is a particularly appropriate supplement to, and extension of, the material of the present text.

A word of caution should be given in connection with the book **Operational Calculus based on the two-sided Laplace integral** by **van der Pol** and **Bremmer** described in the notes at the end of Chapter

2 above. It is worth emphasising once again the value of this classic text for the sheer amount of useful information it contains on transforms and transform methods. But the reader should be warned that the authors use an alternative definition of the Laplace transform once popular among electrical engineers but now rarely to be found. They define the Laplace transform of a function $h(t)$ to be a function which they write as $f(p)$ and which is given by

$$f(p) = p \int_{-\infty}^{+\infty} e^{-pt} h(t) dt.$$

The lack of an appropriate notation for the transform itself, and the use of p rather than s, are features which the reader may find inconvenient, but not otherwise important. But the use of this "p-multiplied" form can be genuinely confusing: it will be especially disconcerting to find that, with this convention, it is the transform of the unit step function which is equal to 1, while that of the delta function is equal to p.

Finally, for a thorough and illuminating discussion of the definition of the Laplace Transform, and more particularly for the debate on the $0+/0-$ problem reference should be made to the paper **Initial conditions, generalized functions, and the Laplace transform** by **K.H.Lundberg, H.R.Miller** and **D.L.Trumper**, (I.E.E.E. Control Systems Magazine, May, 27, 2007, 22-35).

Chapter 6

Fourier Series and Transforms

6.1 FOURIER SERIES

6.1.1 Fourier series; Trigonometric form. The classical theory of the Fourier series expansion of a periodic function can be summarised as follows.

Let f be a real function of t, defined for all t such that $-T/2 < t < +T/2$, and write $\omega_0 \equiv 2\pi/T$. Assume that at all points of the open interval at which f is continuous the following expansion is valid.

$$f(t) = \frac{a_0}{2} + \sum_{n=1}^{+\infty}[a_n \cos(n\omega_0 t) + b_n \sin(n\omega_0 t)]. \qquad (6.1)$$

We can obtain explicit formulas for the coefficients a_0, a_n and b_n appearing in (6.1) by using the so-called **orthogonality relations** for the sine and cosine functions,

$$\int_{-T/2}^{+T/2} \cos(n\omega_0 t)\cos(m\omega_0 t)dt = \begin{cases} T/2 & \text{for } m = n \\ 0 & \text{for } m \neq n \end{cases}$$

$$\int_{-T/2}^{+T/2} \sin(n\omega_0 t)\cos(m\omega_0 t)dt = 0 \qquad \text{for all } m \text{ and } n,$$

$$\int_{-T/2}^{+T/2} \sin(n\omega_0 t)\sin(m\omega_0 t)dt = \begin{cases} T/2 & \text{for } m = n \\ 0 & \text{for } m \neq n \end{cases}$$

Thus, to determine the coefficient a_m for any given integer m, we can multiply (6.1) throughout by $\cos(m\omega_0 t)$ and then integrate from $-T/2$ to $+T/2$. This gives

$$\int_{-T/2}^{+T/2} f(t) \cos(m\omega_0 t)dt =$$

$$\int_{-T/2}^{+T/2} \cos(m\omega_0 t) \left[\frac{a_0}{2} + \sum_{n=1}^{+\infty} a_n \cos(n\omega_0 t) + b_n \sin(n\omega_0 t) \right] dt.$$

Rearranging the terms on the right-hand side, and assuming that it is permissible to integrate the infinite series term by term, we can write this equation in the following form

$$\int_{-T/2}^{+T/2} f(t) \cos(m\omega_0 t)dt =$$

$$\frac{a_0}{2} \int_{-T/2}^{+T/2} \cos(m\omega_0 t)dt + \sum_{n=1}^{+\infty} \int_{-T/2}^{+T/2} \{ a_n \cos(m\omega_0 t) \cos(n\omega_0 t)$$

$$+ b_n \cos(m\omega_0 t) \sin(n\omega_0 t)\} dt$$

Then all the terms on the right-hand side vanish except when $n = m$, and we can solve for the coefficient a_m. Repeating the process with $\sin(m\omega_0 t)$ instead of $\cos(m\omega_0 t)$ we can similarly obtain the coefficient b_m. In the event the required formulas turn out to be

$$a_m = \frac{2}{T} \int_{-T/2}^{+T/2} f(t) \cos(m\omega_0 t)dt,$$

$$b_m = \frac{2}{T} \int_{-T/2}^{+T/2} f(t) \sin(m\omega_0 t)dt$$

for $m = 1, 2, 3, \ldots$, and

$$a_0 = \frac{2}{T} \int_{-T/2}^{+T/2} f(t)dt. \tag{6.2}$$

In point of fact it is not generally true that, for an arbitrary function f, the **Fourier series** given on the right-hand side of (6.1) will converge at each particular point t to the value $f(t)$ assumed by the function f at that point. To ensure that convergence does occur it is necessary to impose extra constraints on f. For our purposes it will be enough to suppose that f satisfies the **Dirichlet conditions** given below, although these are rather more stringent than necessary:

Dirichlet conditions. The function f satisfies the Dirichlet conditions on the interval $(-T/2, +T/2)$ if,

(i) f is bounded on the interval $(-T/2, +T/2)$, and

(ii) the interval $(-T/2, +T/2)$ may be divided into a finite number of sub-intervals in each of which the derivative f' exists throughout and does not change sign.

If these conditions are fulfilled then it is certainly the case that the equation (6.1) is valid at each point of continuity of f in $(-T/2, +T/2)$. Moreover, at any point t at which f is discontinuous, both the one-sided limits $f(t+)$ and $f(t-)$ will necessarily exist and we will have

$$\frac{1}{2}[f(t+) + f(t-)] =$$

$$\frac{a_0}{2} + \sum_{n=1}^{+\infty}[a_n \cos(n\omega_0 t) + b_n \sin(n\omega_0 t)], \qquad (6.3)$$

so that the Fourier series converges to the mean value of the function at such points.

We have assumed so far that f is a function defined only on the fundamental interval $(-T/2, +T/2)$, and have tacitly considered the convergence of the Fourier series of f only at points of this interval. In fact if the series on the right-hand side of (6.1) converges at some point t_0 of $(-T/2, +T/2)$ then it will also converge, and to the same value, at every point of the form $t = t_0 + nT$ where $n = \pm 1, \pm 2, \pm 3, \ldots$. Recalling that we have used the symbol f_T to denote the function obtained by periodically repeating f with period T, we can express this fact by writing

$$f_T(t) \equiv \sum_{n=-\infty}^{+\infty} f(t - nT) \cong \frac{a_0}{2} + \sum_{n=1}^{+\infty}[a_n \cos(n\omega_0 t) + b_n \sin(n\omega_0 t)] \quad (6.4)$$

where the symbol '$\cong$' is here taken to mean equality if t is a point of continuity of f and is to be interpreted in the sense of equation (6.3) otherwise.

6.1.2 Fourier series: Exponential form. It is often convenient to write (6.1) and (6.2) in the following alternative form:

$$f(t) \cong \frac{A_0}{T} + \frac{2}{T} \sum_{n=1}^{+\infty} [A_n \cos(n\omega_0 t) + B_n \sin(n\omega_0 t)] \qquad (6.5)$$

where

$$A_n = \frac{Ta_n}{2} = \int_{-T/2}^{+T/2} f(t) \cos(n\omega_0 t) dt$$

$$B_n = \frac{Tb_n}{2} = \int_{-T/2}^{+T/2} f(t) \sin(n\omega_0 t) dt$$

for $n = 1, 2, 3, \ldots$, and

$$A_0 = \frac{Ta_0}{2} = \int_{-T/2}^{+T/2} f(t) dt.$$

Using these we can easily derive the so-called **exponential form** of the Fourier series of f:

$$f(t) \cong \frac{A_0}{T} + \frac{1}{T} \sum_{n=1}^{+\infty} [A_n(e^{in\omega_0 t} + e^{-in\omega_0 t}) + \frac{B_n}{i}(e^{in\omega_0 t} - e^{-in\omega_0 t})]$$

$$= \frac{A_0}{T} + \frac{1}{T} \sum_{n=1}^{+\infty} [(A_n - iB_n)e^{in\omega_0 t} + (A_n + iB_n)e^{-in\omega_0 t}]$$

$$= \frac{1}{T} \sum_{n=-\infty}^{+\infty} C_n e^{in\omega_0 t} \qquad (6.6)$$

where

$$C_n = A_n - iB_n = \int_{-T/2}^{+T/2} f(t)e^{-in\omega_0 t} dt$$

and

$$C_{-n} = A_n + iB_n = \bar{C}_n. \qquad (6.7)$$

(the bar denoting complex conjugate).

Note that we could have obtained the formulas (6.7) directly by assuming the expansion in the form (6.6) in the first place, multiplying throughout by $\exp(-im\omega_0 t)$ for an arbitrary fixed integer m, and then integrating from $-T/2$ to $+T/2$ to determine C_m.

Exercises I.

1. If f satisfies the Dirichlet conditions in $-T/2 < t < T/2$, and is an even function, show that its Fourier series consists entirely of cosine terms; similarly, if f is an odd function show that its Fourier series consists entirely of sine terms.

2. Obtain Fourier expansions valid in the interval $-\pi < t < +\pi$ for each of the following:

(a) $f_1(t) = t$; (b) $f_2(t) = |t|$; (c) $f_3(t) = t^2$.

By choosing a suitable value of t in each case deduce that

(d) $\pi/4 = 1 - 1/3 + 1/5 - 1/7 + \ldots$,

(e) $\pi^2/8 = 1 + 1/3^2 + 1/5^2 + 1/7^2 + \ldots$,

(f) $\pi^2/12 = 1 - 1/2^2 + 1/3^2 - 1/4^2 + \ldots$.

3. Obtain a Fourier expansion, valid for $\pi < t < +\pi$, for the function $\cos(xt)$ where x is some fixed number which is not an integer. By letting $t \to \pi$ deduce that

$$\cot(\pi x) - \frac{1}{\pi x} = -\frac{2x}{\pi}\left[\frac{1}{1^2 - x^2} + \frac{1}{2^2 - x^2} + \frac{1}{3^2 - x^2} + \cdots\right].$$

4.(a) If $f_1(t) = \cos(t)$ for $0 < t < \pi$, expand f_1 as a Fourier series, valid at least in $0 < t < \pi$, which consists wholly of sine terms. [Hint: extend f_1 as an odd function on $-\pi < t < +\pi$.]

(b) If $f_2(t) = \sin(t)$ for $0 < t < \pi$, expand f_2 as a Fourier series, valid at least in $0 < t < \pi$, which consists wholly of cosine terms. [Hint: extend f_2 as an even function on $-\pi < t < +\pi$,]

6.2   GENERALISED FOURIER SERIES

6.2.1 Periodic impulse train. If we allow a purely formal application of the arguments used in the preceding sections then we can derive very simply a Fourier series representation of a periodic train of delta functions,

$$\sum_{n=-\infty}^{+\infty} \delta(t - nT).$$

Thus, if we asume that

$$\sum_{n=-\infty}^{+\infty} \delta(t - nT) = \frac{1}{T} \sum_{n=-\infty}^{+\infty} C_n e^{inw_0 t}$$

then the coefficients C_n would appear to be given by

$$C_n = \int_{-T/2}^{+T/2} \delta(t) e^{-inw_0 t} dt = \left[e^{-inw_0 t} \right]_{t=0} = 1, \quad \text{for all} \quad n.$$

Hence, again purely formally, we obtain

$$\sum_{n=-\infty}^{+\infty} \delta(t - nT) = \frac{1}{T} \sum_{n=-\infty}^{+\infty} e^{inw_0 t} = \frac{1}{T} \left[1 + 2 \sum_{n=1}^{+\infty} \cos(nw_0 t) \right]. \tag{6.8}$$

The trigonometric and exponential expressions appearing in (6.8) are divergent. This makes it clear that they should be interpreted symbolically, and that the equality signs in (6.8) are to be understood only in the sense that each of the three given expressions has the same operational significance. To be more specific, let $f(t)$ be any continuous function which satisfies the Dirichlet conditions on $(-T/2, +T/2)$ and which vanishes outside that interval. Then

$$\int_{-\infty}^{+\infty} f(t) \left\{ \sum_{n=-\infty}^{+\infty} \delta(t - nT) \right\} dt = f(0)$$

since only the delta function at the origin has any effect on $f(t)$ and the conditions for the sampling operation are certainly satisfied there. Then the series $\frac{1}{T} \sum_{n=-\infty}^{+\infty} e^{inw_0 t}$ is equivalent to $\delta(t)$ on $(-T/2, +T/2)$ in the sense that we are to assign to the expression

$$\int_{-\infty}^{+\infty} f(t) \frac{1}{T} \left\{ \sum_{n=-\infty}^{+\infty} e^{inw_0 t} \right\} dt$$

the value $f(0)$. In the same way we should understand the symbolic expression $\frac{1}{T} \left[1 + 2 \sum_{n=1}^{+\infty} \cos(nw_0 t) \right]$ to behave as an operational equivalent of the delta function $\delta(t)$, at least in so far as the interval $(-T/2, +T/2)$ is concerned.

More generally suppose that $f(t)$ is any continuous function which satisfies the Dirichlet conditions over some finite interval $[a, b]$ and which vanishes identically outside that interval. If m is a positive integer such that

$$[a, b] \subset (-(m+1)T, +(m+1)T)$$

then, from the sampling property of $\delta(t - nT)$, we have

$$\int_{-\infty}^{+\infty} f(t) \left\{ \sum_{n=-\infty}^{+\infty} \delta(t - nT) \right\} dt = \sum_{n=-m}^{+m} f(nT).$$

Accordingly the 'Fourier series' representations of the periodic impulse train given in (6.8) are to be interpreted in the sense that we may write

$$\int_{-\infty}^{+\infty} f(t) \left\{ \frac{1}{T} \sum_{n=-\infty}^{+\infty} e^{in\omega_0 t} \right\} dt = \sum_{n=-m}^{+m} f(nT)$$

$$= \int_{-\infty}^{+\infty} f(t) \left\{ \sum_{n=-\infty}^{+\infty} \delta(t - nT) \right\} dt.$$

And similarly for $\int_{-\infty}^{+\infty} f(t) \frac{1}{T} \left\{ 1 + 2 \sum_{n=1}^{+\infty} \cos(n\omega_0 t) \right\} dt.$

That is to say, each of the formal Fourier series expressions $\frac{1}{T} \sum_{n=-\infty}^{+\infty} e^{in\omega_0 t}$ and $\frac{1}{T} \left[1 + 2 \sum_{n=1}^{+\infty} \cos(n\omega_0 t) \right]$ should really be understood as operationally equivalent to a train of delta functions located at the points $t = nT$, $n = 0, \pm 1, \pm 2, \ldots$.

6.2.2 Remark on types of delta function. In all the work preceding this chapter we have been able to assume that the sampling property associated with a Dirac delta function located at a point a applies to any function f which is continuous at a. However, as remarked earlier in Section 2.5, if we attempt to derive the sampling property for δ by a formal appeal to an integration by parts then it seems necessary to assume that f must be continuously differentiable. In the same way when we consider Fourier representations of delta functions it appears that we must ask for something more than simple continuity for f. This suggests that in certain circumstances it may be appropriate to distinguish between different types of delta function, each type being characterised by some specific variant of the sampling property. Like the problems already noted in connection with products involving delta functions, this is an issue which is difficult to deal with satisfactorily in terms of orthodox classical analysis. Hence we merely note at this stage that a problem does exist and defer further discussion until the introduction of nonstandard analysis in Chapter 10.

6.2.3 Derivation of Fourier coefficients. Some idea of the practical uses to which such symbolic expressions can be put is given by the following illustration of the computation of Fourier coefficients. Consider the derivation of a Fourier series representation for the periodic function shown in Fig. 6.1. This has period $T = 1$, and if we assume that

$$f_T(t) = \frac{1}{T} \sum_{n=-\infty}^{+\infty} C_n e^{i2\pi nt}$$

then the coefficients C_n would be given by the integral formula

$$C_n = \int_{-1/2}^{+1/2} f(t)e^{-i2\pi nt}dt = \int_0^1 (1-t)e^{-i2\pi nt}dt. \qquad (6.9)$$

This integral is not particularly difficult to evaluate, but it is worth noting that we can avoid integration altogether if we appeal to (6.8). We have

$$\frac{d}{dt}f_T(t) = \frac{1}{T} \sum_{n=-\infty}^{+\infty} i2\pi n C_n e^{i2\pi nt} \qquad (6.10)$$

But, as a glance at Fig. 6.1 and Fig. 6.2 shows, the derivative of $f_T(t)$ is given by

$$\frac{d}{dt}f_T(t) = \sum_{n=-\infty}^{+\infty} \delta(t-n) - 1 = \frac{1}{T} \sum_{n=-\infty}^{+\infty} e^{i2\pi nt} - 1 \qquad (6.11)$$

For any $n \neq 0$ we can equate coefficients of $e^{i2\pi nt}$ in (6.10) and (6.11) to obtain

$$i2\pi n C_n = 1, \quad \text{or} \quad C_n = 1/i2\pi n.$$

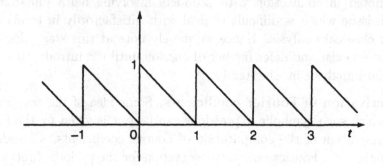

Figure 6.1:

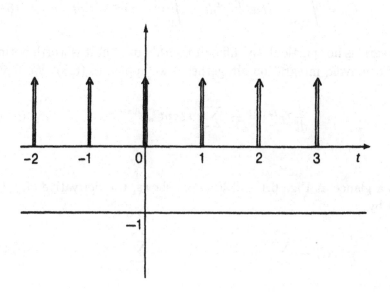

Figure 6.2:

C_n is seen immediately to be $1/2$, since the value of the integral in question is simply the area of the triangle formed by the graph of $f(t)$. Thus, in a particularly simple and straightforward manner we get the Fourier series expansion as

$$f_T(t) \cong \frac{1}{2} + \sum_{n=-\infty, n\neq 0}^{+\infty} \frac{e^{i2\pi nt}}{i2n\pi} = \frac{1}{2} + \sum_{n=1}^{+\infty} \frac{1}{n\pi} \sin(2\pi nt).$$

Exercises II.

1. Let $f(t) = k$ for $-d/2 < t < d/2$, and $f(t) = 0$ otherwise. Express the periodic extension f_T of f, where $T \geq d$, as a Fourier series. Examine the particular cases $d = T$, $d = T/2$.

By taking $k = 1/d$ and allowing d to tend to zero, confirm that the formal Fourier series expansion of a periodic train of delta functions is as given in equation (6.8).

2. Obtain (symbolic) Fourier series for each of the following:

(a) $\displaystyle\sum_{n=-\infty}^{+\infty} \delta(t - 2n\pi)$;   (b) $\displaystyle\sum_{n=-\infty}^{+\infty} \delta[t - (2n + 1)\pi]$;

(c) $\displaystyle\sum_{n=-\infty}^{+\infty} (-1)^n \delta(t - n\pi)$.

3. Show how the results of Question 2 above could be used to derive the Fourier series expansions over $(-\pi, +\pi)$ for the functions f_1, f_2, and f_3 of question 2 of Exercises I.

4. Express $\displaystyle\sum_{n=-\infty}^{+\infty} \delta[t - (2n + 1)\pi/2]$ as a formal Fourier series, and use this result to show that

$$|\cos(t)| = \frac{2}{\pi} + \frac{4}{\pi} \left[\frac{\cos(2t)}{1.3} - \frac{\cos(4t)}{3.5} + \frac{\cos(6t)}{5.7} - \cdots \right]$$

6.3 FOURIER TRANSFORMS

6.3.1 Classical Fourier transform. The full significance and value of the delta function in Fourier analysis can be more readily appreciated if we adopt a somewhat different approach to the study of Fourier series itself. To do so we need first to give a precise definition of the classical **Fourier Transform** of a function.

Let f be a complex-valued function of the real variable t which is **absolutely integrable** over the whole interval $-\infty < t < +\infty$. That is to say we have

$$f(t) = f_1(t) + if_2(t)$$

where

$$\int_{-\infty}^{+\infty} |f_1(t)|dt < +\infty \text{ and } \int_{-\infty}^{+\infty} |f_2(t)|dt < +\infty.$$

Then we define the Fourier transform (strictly, the L^1-**Fourier Transform**) of f to be the function $F(i\omega)$ given by

$$F(i\omega) = \int_{-\infty}^{+\infty} e^{-i\omega t} f(t)dt. \qquad (6.12)$$

The condition of absolute integrability on f is enough to ensure that the function F is well-defined for all ω. In addition it can be shown that it is necessarily bounded and everywhere continuous. We shall sometimes find it convenient to use an operator notation, as with the Laplace transform, and write

$$F(i\omega) \equiv \mathcal{F}[f(t)].$$

$F(i\omega)$ is generally a complex-valued function of ω. If it happens that $f(t)$ itself is a function of t which is wholly real then we have

$$F(i\omega) = \int_{-\infty}^{+\infty} e^{-i\omega t} f(t)dt = A(\omega) - iB(\omega) \qquad (6.13)$$

where the real and imaginary parts of $F(i\omega)$ are given by

$$A(\omega) = \int_{-\infty}^{+\infty} f(t)\cos(\omega t)dt, \quad B(\omega) = \int_{-\infty}^{+\infty} f(t)\sin(\omega t)dt. \qquad (6.14)$$

A is clearly an even function of ω and is often called the **cosine transform** of f; similarly B is an odd function of ω which is often called the **sine transform** of f. Note that the evenness and oddness of the real and imaginary parts of the Fourier transform of f depend essentially on the assumption that f is a real function. If, in particular, f is a real function of t which is itself an even function in t, then we get

$$B(\omega) \equiv 0 \quad \text{and} \quad F(i\omega) = A(\omega) = 2 \int_0^{+\infty} f(t)\cos(\omega t)dt.$$

That is, the Fourier transform of f is wholly real. On the other hand, if f is a real function of t which is odd in t, then

$$A(\omega) \equiv 0 \quad \text{and} \quad F(i\omega) = -iB(\omega) = -2i \int_0^{+\infty} f(t)\sin(\omega t)dt,$$

so that the Fourier transform of f is purely imaginary.

6.3.2 Fourier transforms and Fourier series. Suppose now that f is a (real) function which vanishes identically outside the finite interval $-T/2 < t < +T/2$ and which satisfies the Dirichlet conditions within that interval. Then

$$F(i\omega) = \int_{-\infty}^{+\infty} f(t)e^{-i\omega t}dt = \int_{-T/2}^{+T/2} f(t)e^{-i\omega t}dt$$

and

$$f_T(t) \cong \frac{A_0}{T} + \frac{2}{T}\sum_{n=1}^{+\infty}[A_n\cos(n\omega_0 t) + B_n\sin(n\omega_0 t)]$$

$$= \frac{1}{T}\sum_{n=-\infty}^{+\infty} C_n e^{in\omega_0 t}$$

where

$$C_n = \int_{-T/2}^{+T/2} f(t)e^{-in\omega_0 t}dt \equiv F(in\omega_0)$$

$$A_n = \int_{-T/2}^{+T/2} f(t)\cos(n\omega_0 t)dt \equiv A(n\omega_0) \qquad (6.15)$$

$$B_n = \int_{-T/2}^{+T/2} f(t)\sin(n\omega_0 t)dt \equiv B(n\omega_0).$$

To illustrate this relationship between Fourier transforms and Fourier series, consider the simple but important example afforded by the rectangular pulse shown in Fig. 6.3(a).

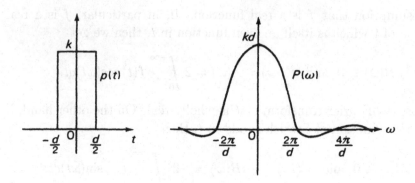

Figure 6.3:

The corresponding Fourier transform, shown in Fig. 6.3(b), is easily computed as follows:

$$P(i\omega) = \int_{-\infty}^{+\infty} p(t)e^{-i\omega t}dt = k \int_{-d/2}^{+d/2} e^{-i\omega t}dt = \frac{2k}{\omega}\sin(\omega d/2)$$

Hence we have

$$P(i\omega) = kd\left[\frac{\sin(\omega d/2)}{\omega d/2}\right]. \tag{6.16}$$

If T is any number greater than or equal to d then we can confirm the result of Question 3 of Exercises I by writing down at once the Fourier series for the function p_T obtained by periodically repeating p with period T (see Fig. 6.4):

$$p_T(t) \equiv \sum_{n=-\infty}^{+\infty} p(t-nT) \cong \frac{1}{T}\sum_{n=-\infty}^{+\infty} P(in\omega_0)e^{in\omega_0 t}$$

$$= \frac{kd}{T}\sum_{n=-\infty}^{+\infty}\left[\frac{\sin(n\omega_0 d/2)}{n\omega_0 d/2}\right]e^{in\omega_0 t}$$

$$= \frac{kd}{T}\left[1 + 2\sum_{n=1}^{+\infty}\frac{\sin(n\omega_0 d/2)}{n\omega_0 d/2}\cos(n\omega_0 t)\right]. \tag{6.17}$$

The Fourier coefficients C_n are given by the ordinates $P(in\omega_0)$ as shown in Fig. 6.5). With the usual interpretation of t as time and ω

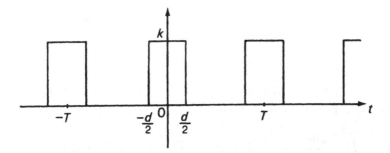

Figure 6.4:

as (angular) frequency we might sum this up by saying that periodic extension of a function in the time domain corresponds to periodic sampling in the frequency domain.

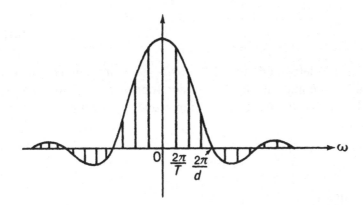

Figure 6.5:

6.3.3 Fourier Inversion Theorem.

There is an **Inversion theorem** for the Fourier transform which is of crucial importance in what follows. A rigorous proof is beyond the scope of this text, and we give here only an elementary, non-rigorous derivation of it. Nevertheless the argument does show how a judicious use of the delta function can make the result simple and conceptually clear.

We use the standard result (obtainable, for example, by simple contour integration)

$$P \int_{-\infty}^{+\infty} \frac{e^{imx}}{x} dx = \pi i \qquad (6.18)$$

where the symbol P denotes the Cauchy Principal Value of the integral and m is a positive constant. That is,

$$\int_{-\infty}^{+\infty} \frac{\sin(mx)}{x} dx = \pi \quad \text{and} \quad \int_{-\infty}^{+\infty} \frac{\cos(mx)}{x} dx = 0$$

Replacing m by $-m$ simply changes the sign of the first of these two real integrals and leaves the other unaltered. Hence, if we replace m by the usual symbol t for the independent real variable we can write

$$\frac{1}{2\pi} \int_{-\infty}^{+\infty} \frac{e^{itx}}{ix} dx = \frac{1}{2} \operatorname{sgn}(t) \equiv \begin{cases} +1/2 & \text{for } t > 0 \\ -1/2 & \text{for } t < 0 \end{cases}$$

A formal differentiation of this with respect to t then yields

$$\frac{1}{2\pi} \int_{-\infty}^{+\infty} e^{itx} dx = \delta(t). \qquad (6.19)$$

The integral on the left-hand side of (6.19) is, of course, divergent, and it is clear that this equation must be understood symbolically. That is to say, for all sufficiently well-behaved functions f, we should interpret (6.19) to mean that

$$\int_{-\infty}^{+\infty} f(t) \left[\frac{1}{2\pi} \int_{-\infty}^{+\infty} e^{i\omega t} d\omega \right] dt = \int_{-\infty}^{+\infty} f(t)\delta(t) dt = f(0)$$

or, more generally, that

$$\int_{-\infty}^{+\infty} f(\tau) \left[\frac{1}{2\pi} \int_{-\infty}^{+\infty} e^{i(t-\tau)\omega} d\omega \right] d\tau$$

$$= \int_{-\infty}^{+\infty} f(\tau)\delta(t-\tau) d\tau = f(t). \qquad (6.20)$$

With due disregard for any of the niceties involved in altering the order of the integration signs appearing in (6.20) we can rewrite this result in the form

$$f(t) = \frac{1}{2\pi} \int_{-\infty}^{+\infty} e^{i\omega t} \left[\int_{-\infty}^{+\infty} f(\tau) e^{-i\omega \tau} d\tau \right] d\omega$$

so that

$$f(t) = \frac{1}{2\pi} \int_{-\infty}^{+\infty} e^{i\omega t} F(i\omega) d\omega \qquad (6.21)$$

where $F(i\omega)$ denotes, as usual, the Fourier transform of f. Thus it appears that given the function $F(i\omega)$ we can recover from it the function $f(t)$ of which it is the Fourier transform. The validity of the inversion formula (6.21) will obviously depend critically on the conditions which must be imposed on the function f to make the above assertions and manipulations justifiable. (This is of course the crux of every genuine proof of the result and is precisely what we have glossed over above by using the phrase 'sufficiently well- behaved'.) For completeness we state a form of the **Fourier Inversion Theorem** proper which is sufficiently comprehensive for most purposes and which can be established rigorously.

Fourier Inversion Theorem. Let f be a (real or complex valued) function of a single real variable which is absolutely integrable over the interval $(-\infty, +\infty)$ and which also satisfies the Dirichlet conditions over every finite interval. If $F(i\omega)$ denotes the Fourier transform of f then at each point t we have

$$\frac{1}{2\pi} \int_{-\infty}^{+\infty} e^{i\omega t} F(i\omega) d\omega = \frac{1}{2}[f(t+) + f(t-)]. \qquad (6.22)$$

(At all points of continuity of f, (6.22) reduces to (6.21).)

6.3.4 Properties of the Fourier transform. As in the case of the Laplace transform there are several important properties of the Fourier transform which merit explicit mention. It is a simple matter to establish the following (see Exercises III, Question 2):

(i) **Linearity:** $\mathcal{F}[af_1(t) + bf_2(t)](i\omega) = aF_1(i\omega) + bF_2(i\omega)$,

(ii) $\mathcal{F}[f(t-a)](i\omega) = \exp(-i\omega a)F(i\omega)$,

(iii) $\mathcal{F}[f(t)\exp(-at)](i\omega) = F(i\omega + a) \equiv F(i(\omega - ia))$,

(iv) $\mathcal{F}[f'(t)](i\omega) = i\omega F(i\omega)$.

Moreover there is a convolution theorem for the Fourier transform which compares directly with that for the Laplace transform: Let

$$h(t) = (f \star g)(t) = \int_{-\infty}^{+\infty} f(t - \tau)g(\tau)d\tau.$$

Then

$$H(i\omega) = \int_{-\infty}^{+\infty} e^{-i\omega t} dt \int_{-\infty}^{+\infty} f(t-\tau)g(\tau)d\tau$$

$$= \int_{-\infty}^{+\infty} g(\tau)d\tau \int_{-\infty}^{+\infty} f(t-\tau)e^{-i\omega t} dt$$

$$= \int_{-\infty}^{+\infty} g(\tau)d\tau \int_{-\infty}^{+\infty} f(x)e^{-i\omega(x+\tau)} dx$$

$$= \int_{-\infty}^{+\infty} e^{-i\omega\tau} g(\tau)d\tau \int_{-\infty}^{+\infty} e^{-i\omega x} f(x)dx = F(i\omega)G(i\omega).$$

Thus the Fourier transform of the convolution of two functions is equal to the product of their individual transforms.

Using the Fourier Inversion Theorem we can derive a dual result which relates to the convolution of Fourier transforms themselves. Thus, given that $F(i\omega) \equiv \mathcal{F}[f(t)]$ and that $G(i\omega) \equiv \mathcal{F}[g(t)]$ are themselves absolutely integrable functions over $(-\infty, +\infty)$, consider their convolution

$$H(i\omega) = \int_{-\infty}^{+\infty} F[i(\omega - \nu)]G(i\nu)d\nu.$$

From the inversion Theorem,

$$\mathcal{F}^{-1}[H(i\omega)] \equiv h(t) = \frac{1}{2\pi} \int_{-\infty}^{+\infty} H(i\omega)e^{i\omega t} d\omega$$

$$= \frac{1}{2\pi} \int_{-\infty}^{+\infty} e^{i\omega t} d\omega \int_{-\infty}^{+\infty} F[i(\omega - \nu)]G(i\nu)d\nu$$

$$= \frac{1}{2\pi} \int_{-\infty}^{+\infty} G(i\nu)d\nu \int_{-\infty}^{+\infty} e^{i\omega t} F[i(\omega - \nu)]d\omega$$

$$= \frac{1}{2\pi} \int_{-\infty}^{+\infty} G(i\nu)d\nu \int_{-\infty}^{+\infty} F(i\mu)e^{i(\mu+\nu)t} d\mu$$

$$= \frac{1}{2\pi} \int_{-\infty}^{+\infty} G(i\nu)e^{i\nu t} d\nu \int_{-\infty}^{+\infty} F(i\mu)e^{i\mu t} d\mu = g(t)[2\pi f(t)].$$

Accordingly we may state the following dual results for the Fourier transforms of convolutions and products,

$$\mathcal{F}[(f(t) \star g(t)] = F(i\omega)G(i\omega), \qquad (6.23a)$$

$$\mathcal{F}[f(t)g(t)] = \frac{1}{2\pi}[F(i\omega) \star G(i\omega)]. \tag{6.23b}$$

Exercises III.

1. If $\mathcal{F}[f(t)] = F(i\omega)$ show that

(a) $\mathcal{F}[f(t-a)] = \exp(-i\omega a)F(i\omega)$,

(b) $\mathcal{F}[f(t)\exp(-at)] = F(i\omega + a)$,

(c) $\mathcal{F}[f(at)] = \frac{1}{a}F(i\omega/a)$,

(d) $\mathcal{F}[f'(t)] = i\omega F(i\omega)$,

where in (d) we assume that the classical derivative of f exists and is an absolutely integrable function over $(-\infty, +\infty)$.

2. If $h_1(t) = 1$ for $|t| < a/2$ and $= 0$ otherwise, and if $h_2(t) = 1$ for $|t| < b/2$ and $= 0$ otherwise, find the function $h_3(t) = h_1(t) \star h_2(t)$, and confirm by direct calculation that

$$H_3(i\omega) = H_1(i\omega)H_2(i\omega).$$

3. Find the Fourier transforms of each of the following functions:

(a) $\exp(-|t|)$; (b) $\text{sgn}(t)\exp(-|t|)$; (c) $\exp(-t^2/2)$;

(d) $f_1(t) = 1 - t^2$ for $|t| < 1$, and $= 0$ otherwise;

(e) $f_2(t) = \sin(t)$ for $0 < t < \pi$, and $= 0$ otherwise;

(f) $f_3(t) = \cos(t)$ for $-\pi/2 < t < +\pi/2$, and $= 0$ otherwise.

[Hint: in (c) recall the fact that $\int_{-\infty}^{+\infty} e^{-x^2} dx = \sqrt{\pi}$ and use the equation $t^2 + 2at = (t+a)^2 - a^2$].

4. Let f be a real function of t which is absolutely integrable over $(-\infty, +\infty)$ and continuous everywhere. Use the Fourier Inversion Theorem to show that

$$f(t) = \frac{1}{\pi} \int_0^{+\infty} d\omega \int_{-\infty}^{+\infty} f(\tau)\cos\omega(t-\tau)d\tau.$$

Deduce that, if $f(t)$ is even, then

$$f(t) = \frac{2}{\pi} \int_0^{+\infty} \cos(\omega t)d\omega \int_0^{+\infty} f(\tau)\cos(\omega\tau)d\tau,$$

while, if $f(t)$ is odd, then

$$f(t) = \frac{2}{\pi} \int_0^{+\infty} \sin(\omega t) d\omega \int_0^{+\infty} f(\tau) \sin(\omega \tau) d\tau.$$

5. By using the results of (3) and (4) above evaluate the integrals

(a) $\displaystyle\int_0^{+\infty} \frac{\cos(xt)}{x^2 + 1} dx;$ ' (b) $\displaystyle\int_0^{+\infty} \left[\frac{\omega \cos(\omega) - \sin(\omega)}{\omega^3} \right] d\omega.$

6.4   GENERALISED FOURIER TRANS-FORMS

6.4.1 The L^2 Fourier transform. A function $n(t)$ is said to be a **null-function** on the interval $-\infty < t < +\infty$ if

$$\int_{-\infty}^{+\infty} |n(t)| dt = 0.$$

Two functions which differ only by a null function are said to be equal **almost everywhere**, and we write $f_1(t) =_{a.e.} f_2(t)$. This means, in effect, that the set of points at which the two functions differ in value is a negligibly small set in the sense that it has zero total length. We have already used this idea in connection with Laplace transform inversion (Sec. 5.1.4), and the point will be taken up again in Chapter 8. In the present context we can use it to give an alternative statement of the Fourier Inversion Theorem.

Let f be absolutely integrable over $-\infty < t < +\infty$. The Fourier transform of f, given for all ω by

$$F(i\omega) = \int_{-\infty}^{+\infty} f(t) e^{-i\omega t} dt, \tag{6.24}$$

is a bounded, continuous function of ω which is such that

$$\lim_{|\omega| \to \infty} F(i\omega) = 0.$$

But $F(i\omega)$ is not generally a function which is itself absolutely integrable over $-\infty < \omega < +\infty$. Hence, the inversion formula ought properly to be expressed in the form of a limit,

$$f(t) =_{a.e.} \lim_{m \to \infty} \frac{1}{2\pi} \int_{-m}^{+m} F(i\omega)e^{i\omega t}d\omega. \qquad (6.25)$$

since the convergence as $m \to \infty$ may only be conditional. Secondly we know that at points of discontinuity of f the right-hand side of (6.25) may fail to converge to the values assumed by the function: the use of the symbol '$=_{a.e.}$' indicates that the totality of such points is negligibly small.

Now for many purposes we need to work with functions $f(t)$ which are not necessarily absolutely integrable over $(-\infty, +\infty)$, and for which the classical Fourier transform of (6.24) is not defined. Suppose first that $f(t)$ is **square-integrable** over $(-\infty, +\infty)$; that is to say, that we have

$$\int_{-\infty}^{+\infty} |f(t)|^2 dt < +\infty.$$

Then $f(t)$, which may or may not be absolutely integrable over $(-\infty, +\infty)$, is often described as a 'function of finite energy'. In such a case, for each $m \in \mathbb{N}$, we can define the function $f_m(t)$ as equal to $f(t)$ on $(-m, +m)$ and equal to zero elsewhere. Each such function f_m is absolutely integrable over $(-\infty, +\infty)$ and therefore has a well-defined Fourier transform given by

$$F_m(i\omega) \equiv \int_{-m}^{+m} f(t)e^{-i\omega t}dt$$

For each m, $F_m(i\omega)$ is itself a function which is square-integrable over $(-\infty, +\infty)$. Further, it can be shown that there always exists a square-integrable function $F(i\omega)$ such that

$$\lim_{m \to \infty} \int_{-\infty}^{+\infty} |F(i\omega) - F_m(i\omega)|^2 d\omega = 0. \qquad (6.26)$$

This function we now define to be the so-called L^2 **Fourier transform** of the original square-integrable function $f(t)$. Strictly speaking we should say that the L^2-transform of f is not a single function but an equivalence class of square-integrable functions which are equal almost everywhere.

However, suppose further that $f(t)$ is both absolutely integrable and square-integrable over $(-\infty, +\infty)$. Then it already has a well-defined Fourier transform $F(i\omega)$ which is a continuous function of ω, and this itself satisfies the equation (6.26). Hence the new definition of Fourier transform is compatible with the original definition of the classical L^1-Fourier transform in (6.24) whenever the latter exists. But in general if f is merely square-integrable over $(-\infty, +\infty)$ then $F(i\omega)$ need not be continuous, nor even bounded. Nevertheless it does turn out to be always a square-integrable function itself over $-\infty < \omega < +\infty$, so that this extended L^2-transform maps functions of finite energy into functions of finite energy.

For example, the function $f(t) = \frac{1}{\pi t}\sin(\pi t)$ is not absolutely integrable over $(-\infty, +\infty)$, and so it does not have a uniquely well-defined L^1-transform given by equation (6.24). On the other hand we do have $\int_{-\infty}^{+\infty} |f(t)|^2 dt = 1$. Hence this function is square-integrable over $(-\infty, +\infty)$ and so it does have a Fourier transform in the L^2 sense. A function $F(i\omega)$ which behaves in the required way for this can be computed as the limit

$$F(i\omega) = \lim_{m\to\infty} \int_{-m}^{+m} \frac{\sin(\pi t)}{\pi t} e^{-i\omega t} dt = \begin{cases} 1 & \text{for } |\omega| < \pi \\ 0 & \text{for } |\omega| > \pi \end{cases}$$

This L^2-Fourier transform is a function which is plainly discontinuous at $\omega = \pm\pi$. Moreover we could complete the definition of $F(i\omega)$ in any way we wish at those points of discontinuity, to give a whole family of functions equal almost everywhere, any one of which could serve as a representative of the required Fourier transform.

In spite of this extension of the Fourier transform to functions of finite energy, there still remains a variety of important and useful functions which are neither absolutely integrable nor square-integrable, and it is desirable to extend the meaning of Fourier transform yet again so as to apply to these. It is this which requires us once more to turn to the use of generalised functions.

6.4.2 Fourier transforms of delta functions. Applying the sampling property of the delta function to the case of the Fourier integral itself we get

$$\int_{-\infty}^{+\infty} e^{-i\omega t}\delta(t)dt = 1 \qquad (6.27)$$

or, more generally,

$$\int_{-\infty}^{+\infty} e^{-i\omega t}\delta(t-a)dt = e^{-i\omega a}. \tag{6.28}$$

This suggests that we define the Fourier transform of $\delta(t-a)$ to be $\exp(-i\omega a)$. The inversion formula would then be satisfied formally since, using (6.19), we would have

$$\frac{1}{2\pi}\int_{-\infty}^{+\infty} e^{i\omega t}e^{-i\omega a}d\omega = \frac{1}{2\pi}\int_{-\infty}^{+\infty} e^{i\omega(t-a)}d\omega = \delta(t-a).$$

Dually we can apply the sampling property of the delta function to the Fourier inversion integral:

$$\frac{1}{2\pi}\int_{-\infty}^{+\infty} e^{i\omega t}\delta(\omega-\alpha)d\omega = \frac{1}{2\pi}e^{i\alpha t}$$

and similarly

$$\frac{1}{2\pi}\int_{-\infty}^{+\infty} e^{i\omega t}\delta(\omega+\alpha)d\omega = \frac{1}{2\pi}e^{-i\alpha t}.$$

Thus, recalling that the Fourier transform is defined in general for complex-valued functions, these results suggest that we can give the following definitions for the Fourier transforms of complex exponentials such as $e^{i\alpha t}$ and $e^{-i\alpha t}$:

$$\mathcal{F}[e^{i\alpha t}] = 2\pi\delta(\omega-\alpha) \quad ; \quad \mathcal{F}[e^{-i\alpha t}] = 2\pi\delta(\omega+\alpha). \tag{6.29}$$

These equations (6.29) immediately yield the following definitions for the Fourier transforms of the real functions $\cos(\alpha t)$ and $\sin(\alpha t)$:

$$\mathcal{F}[\cos(\alpha t)] = \pi[\delta(\omega-\alpha) + \delta(\omega+\alpha)]$$

$$\mathcal{F}[\sin(\alpha t)] = \frac{\pi}{i}[\delta(\omega-\alpha) - \delta(\omega+\alpha)]. \tag{6.30}$$

In particular, taking $\alpha = 0$, we find that the generalised Fourier transform of the constant function $f(t) \equiv 1$ is simply $2\pi\delta(\omega)$. This in turn allows us to offer a definition of the Fourier transform of the unit step function. For we have

$$u(t) = \frac{1}{2} + \frac{1}{2}\,\mathrm{sgn}(t)$$

and from (6.18) we know that $2/i\omega$ is a suitable choice for the Fourier transform of the function $\text{sgn}(t)$ in the sense that

$$\text{sgn}(t) = \frac{1}{2\pi} \int_{-\infty}^{+\infty} \frac{2}{i\omega} e^{i\omega t} d\omega.$$

Accordingly it follows that we should set

$$\mathcal{F}[u(t)] = \pi\delta(\omega) + \frac{1}{i\omega}. \tag{6.31}$$

Finally, recall that we have already established some formal 'Fourier series' expansions for periodic trains of delta functions:

$$\sum_{n=-\infty}^{+\infty} \delta(t-nT) = \frac{1}{T}\left[1 + 2\sum_{n=1}^{+\infty}\cos(n\omega_0 t)\right] = \frac{1}{T}\sum_{n=-\infty}^{+\infty} e^{in\omega_0 t}. \tag{6.32}$$

We now examine the possibility of extending the definition of Fourier transform to apply to these symbolic expressions. First, since the Fourier Transform of $\delta(t-nT)$ is $e^{-inT\omega}$, we ought to get

$$\mathcal{F}\left[\sum_{n=-\infty}^{+\infty}\delta(t-nT)\right] = \sum_{n=-\infty}^{+\infty} e^{-inT\omega} \equiv \sum_{n=-\infty}^{+\infty} e^{inT\omega}. \tag{6.33}$$

On the other hand, the Fourier transform of $e^{in\omega_0 t}$ has been computed as $2\pi\delta(\omega-n\omega_0)$, and similarly that of $\cos(n\omega_0 t)$ as $\pi[\delta(\omega-n\omega_0)+\delta(\omega+n\omega_0)]$. Accordingly the Fourier transform of the exponential and trigonometric expressions appearing in (6.32) would appear to be

$$\mathcal{F}\left[\frac{1}{T}\{1 + 2\sum_{n=1}^{+\infty}\cos(n\omega_0 t)\}\right] = \mathcal{F}\left[\frac{1}{T}\sum_{n=-\infty}^{+\infty} e^{in\omega_0 t}\right]$$

$$= \omega_0 \sum_{n=-\infty}^{+\infty} \delta(\omega - n\omega_0). \tag{6.34}$$

Thus we obtain the very important result that the (generalised) Fourier transform of a periodic train of delta functions in the time domain is a corresponding periodic train of delta functions in the frequency domain:

$$\mathcal{F}\left[\sum_{n=-\infty}^{+\infty}\delta(t-nT)\right] = \frac{2\pi}{T}\sum_{n=\infty}^{+\infty} \delta(\omega - n2\pi/T), \tag{6.35}$$

where, for the sake of clarity, we replace ω_0 by $2\pi/T$. Further, comparing (6.33) and (6.34), we get

$$\sum_{n=-\infty}^{+\infty} e^{in\omega T} = \frac{2\pi}{T} \sum_{n=-\infty}^{+\infty} \delta(\omega - n2\pi/T) \tag{6.36}$$

which may again be interpreted as a (formal) Fourier series representation of the periodic impulse train in the frequency domain.

6.4.3 Fourier transforms of periodic functions. We shall now apply the considerations of the preceding section to derive the generalised Fourier transform of an arbitrary periodic function. Suppose that f is a bounded, continuous function of t which vanishes outside the finite interval $(-T/2, +T/2)$ and which satisfies the Dirichlet conditions within that interval. Using the fact that the Fourier transform of $f(t - nT)$ is $e^{-inT\omega}F(i\omega)$, we get

$$\mathcal{F}\left[\sum_{n=-\infty}^{+\infty} f(t-nT)\right] = \sum_{n=-\infty}^{+\infty} F(i\omega)e^{-inT\omega}$$

$$= F(i\omega) \sum_{n=-\infty}^{+\infty} e^{-inT\omega} = \omega_0 \sum_{n=-\infty}^{+\infty} F(i\omega)\delta(\omega - n\omega_0)$$

$$= \frac{2\pi}{T} \sum_{n=-\infty}^{+\infty} F(i2\pi n/T)\delta(\omega - 2\pi n/T).$$

Now formally apply the inversion formula to this last expression,

$$\sum_{n=-\infty}^{+\infty} f(t-nT) \cong$$

$$\frac{1}{2\pi} \int_{-\infty}^{+\infty} e^{i\omega t}\left[\frac{2\pi}{T} \sum_{n=-\infty}^{+\infty} F(2\pi ni/T)\delta(\omega - 2\pi n/T)\right] d\omega$$

$$= \frac{1}{T} \sum_{n=-\infty}^{+\infty} F(2\pi ni/T)e^{in2\pi t/T}$$

and we recover the usual Fourier series expansion of the periodic extension of f.

This result does support the alleged formal equivalence of the expressions in (6.32). For we know that

$$\sum_{n=-\infty}^{+\infty} f(t-nT) = f(t) \star \sum_{n=-\infty}^{+\infty} \delta(t-nT)$$

and so we ought to expect that the Fourier transform of the right-hand side should be the product

$$F(i\omega) \times \hat{F}\left[\sum_{n=-\infty}^{+\infty} \delta(t-nT)\right].$$

That this is the case follows precisely because of the derivation of the Fourier transform of the impulse train from the 'Fourier series' expansions of (6.32).

Exercises IV.

1. Find the (generalised) Fourier transforms of each of the following:

(a) $f_1(t) = (1 - e^{-at})u(t)$;   (b) $f_2(t) = \cos^2(at)$;

(c) $f_3(t) = \sin^2(at)$.

2. Delta functions are located at the points

$$t = 0, T, 2T, \ldots, (2N-1)T.$$

Calculate the Fourier transform of this finite train of delta functions and hence find the inverse Fourier transform of

$$\frac{\sin(N\omega T)}{\sin(\omega T/2)}.$$

3. A real-valued, continuous function f, with Fourier transform $F(i\omega)$, is periodically sampled at the points $t = nT$ (where $n = 0, \pm 1, \pm 2, \ldots$). This may be regarded as an operation which transforms $f(t)$ into the generalised function

$$f_{(*)}(t) \equiv \sum_{n=-\infty}^{+\infty} f(nT)\delta(t-nT)$$

by means of the formal multiplication

$$f_{(*)}(t) \equiv \sum_{-\infty}^{+\infty} f(nT)\delta(t - nT) = f(t) \times \sum_{n=-\infty}^{+\infty} \delta(t - nT).$$

Assuming that the infinite series concerned do converge, show formally that

$$\mathcal{F}[f_{(*)}(t)] = \frac{1}{T} \sum_{n=-\infty}^{+\infty} F[i(\omega - 2\pi n/T)] = \sum_{n=-\infty}^{+\infty} f(nT)e^{-in\omega T}.$$

4. Suppose that in Question 3 the Fourier transform $F(i\omega)$ of the function f vanishes identically outside the interval $(\pi/T, +\pi/T)$. Let $h(t)$ be the impulse response function of a time-invariant linear system whose transfer function, $H(i\omega)$, is equal to 1 for all ω such that $|\omega| < \pi T$ and is equal to 0 otherwise. (Such a T.I.L.S. is often called an **ideal low-pass filter**.) Find the output signal which results when the sampled function $f_{(*)}$ is applied as an input, and hence show that

$$f(t) = \sum_{n=-\infty}^{+\infty} f(nT) \left[\frac{\sin(t - nT)\pi/T}{(t - nT)\pi/T} \right].$$

Further reading.

A concise rigorous account of account of the fundamental theory of the classical Fourier transform is given by **R.R.Goldberg** in **Fourier Transforms**, published as one of the Cambridge University Tracts in Mathematics and Mathematical Physics, (C.U.P., 1961). This does require some basic acquaintance with the Lebesgue integral (outlined in Chapter 8 of the present text), but is a very clear and readable book. For a lively and illuminating introduction to generalised Fourier transforms and their applications the reader is once again very strongly recommended to turn to the book by **Sir James Lighthill** already cited in preceding chapters.

Chapter 7

Other Generalised Functions

7.1 FRACTIONAL CALCULUS

7.1.1 A Cauchy Principal Value. Recall once more that the term 'generalised function' refers to a mathematical entity which is not strictly a function at all in the proper sense of the word, but is defined in terms of its action on other, bona fide functions. In the first part of this book we have been concerned almost exclusively with the unit step function, the delta function and derivatives of the delta function, and with the formal rules which should ensure correct usage of them. However, the delta function and its derivatives are not the only generalised functions which are of importance in applied analysis although they are certainly the most well known. Consider, for example, the case of the ordinary function $1/t$. This is defined for all $t \in \mathbb{R}$ except the origin, in the neighbourhood of which it becomes unbounded in absolute value. Moreover it is not an absolutely integrable function over $(-\infty, +\infty)$ and is not even locally integrable over any interval which includes the origin. It follows that if $f(t)$ is an arbitrary continuous function on $\mathbb{R}$ which vanishes outside some finite interval, then the integral

$$\int_{-\infty}^{+\infty} \frac{f(t)}{t} dt \tag{7.1}$$

will not generally converge, even though the integration is in effect only over a finite range. We could, however, obtain a specific finite value from this expression by computing it as a Cauchy Principal Value:

$$P \int_{-\infty}^{+\infty} \frac{f(t)}{t} dt = \lim_{\epsilon \downarrow 0} \left\{ \int_{-\infty}^{-\epsilon} \frac{f(t)}{t} dt + \int_{-\epsilon}^{+\infty} \frac{f(t)}{t} dt \right\}. \tag{7.2}$$

154

In this way we would have defined an operation which maps each continuous function f which vanishes outside some finite interval into a certain real number. The parallel with the defining characteristic of the delta function (viz. the sampling property) is obvious, and one is tempted to use a similar notation and express this operation in the following form:

$$f \to < P\left(\frac{1}{t}\right), f >= P \int_{-\infty}^{+\infty} \frac{f(t)}{t} dt. \tag{7.3}$$

The use of the notation $P\left(\frac{1}{t}\right)$ in (7.3) is deliberate. The value which we have denoted by $< P\left(\frac{1}{t}\right), f >$ has been computed not by carrying out a genuine integration process using the ordinary function $1/t$, but by means of a special device (the Cauchy Principal Value) which extracts a finite quantity from an otherwise divergent integral. Just as $\delta(t)$ needs to be distinguished from the ordinary function $u'(t)$ which is the classical derivative of $u(t)$, so $P\left(\frac{1}{t}\right)$ needs to be distinguished from the ordinary function $1/t$ which is the classical derivative of $\log(|t|)$. It is another example of a generalised function which, as will be seen below, behaves as the generalised derivative of $\log(|t|)$.

Sooner or later it becomes necessary to develop a systematic and comprehensive theory of all such generalised functions and we shall sketch briefly how this may be done later on in the next chapter. Before doing so however it will be useful to give some more specific examples of generalised functions other than delta functions, and to indicate a context in which they may be seen to be significant. In order to do this we will need first to introduce another important standard function which will be useful in the sequel.

7.1.2 The Gamma Function. For all $\alpha > 0$ the so-called **Gamma Function**, $\Gamma(\alpha)$, is defined by the integral

$$\Gamma(\alpha) = \int_0^{+\infty} t^{\alpha-1} e^{-t} dt. \tag{7.4}$$

That this integral does converge for all such α is easily confirmed as follows. First, for any $a > 0$, we can write

$$\Gamma(\alpha) = \int_0^a t^{\alpha-1} e^{-t} dt + \int_a^{+\infty} t^{\alpha-1} e^{-t} dt. \tag{7.5}$$

Near $t = 0$ the function $t^{\alpha-1} e^{-t}$ behaves like $t^{\alpha-1}$ and so the first integral on the right-hand side of (7.5) is convergent. Also, if $t > 0$ then for any positive integer p we have

$$e^{-t} < p!/t^p \quad \text{and so} \quad e^{-t}t^{\alpha-1} < p!/t^{p-\alpha+1}.$$

Hence

$$\int_a^{+\infty} t^{\alpha-1}e^{-t}dt < p! \int_a^{+\infty} \frac{dt}{t^{p-\alpha+1}},$$

and we have only to choose for p any positive integer greater than α to confirm that the second integral on the right-hand side of (7.5) also converges for any $\alpha > 0$, and therefore that the definition (7.4) does make sense for all such values of α.

The most important characteristic property of the gamma function is the following:

$$\Gamma(\alpha + 1) = \alpha\Gamma(\alpha). \tag{7.6}$$

This can be established by a simple integration by parts. For we have

$$\Gamma(\alpha + 1) = \int_0^{+\infty} t^\alpha e^{-t}dt$$

$$= \left[-t^\alpha e^{-t}\right]_0^{+\infty} + \alpha \int_0^{+\infty} t^{\alpha-1}e^{-t}dt = \alpha\Gamma(\alpha).$$

In particular, since $\Gamma(1) = \int_0^{+\infty} e^{-t}dt = 1$, it follows that whenever we take for α a positive integer, n, we get

$$\Gamma(n+1) = n\Gamma(n) = n(n-1)\Gamma(n-1) = \ldots$$

$$= n(n-1)(n-2)\ldots 3.2.1.\Gamma(1) = n! \tag{7.7}$$

Hence an immediate consequence of the characteristic relation (7.6) of the gamma function is that the domain of definition of the factorial function can be extended from the positive integers to the whole positive real axis. Thus, for example,

$$\Gamma(1/2) = \int_0^{+\infty} t^{-1/2}e^{-t}dt = 2\int_0^{+\infty} e^{-\tau^2}d\tau = \sqrt{\pi}$$

so that we can use (7.6) to define

$$(1/2)! \equiv \Gamma(1+1/2) = \frac{1}{2}\Gamma(1/2) = \sqrt{\pi}/2.$$

We can even assign significance to such apparently meaningless expressions as 0! since we may write

$$0! \equiv \Gamma(1) = 1. \tag{7.8}$$

7.1.3 Fractional integration. Rather more significantly we can use the gamma function to extend the meaning of the integration operator. It will be convenient to adopt the following systematic notation:

for any real $\lambda > 0$ the symbol $t_+^{\lambda-1}$ will denote the function given by

$$t_+^{\lambda-1}(t) \equiv |t|^{\lambda-1} u_0(t) = \begin{cases} |t|^{\lambda-1} & \text{for } t > 0 \\ 0 & \text{for } t < 0 \end{cases} \tag{7.1}$$

For brevity in what follows we often write $t^{\lambda-1}u(t)$ instead of the more correct expression $|t|^{\lambda-1}u_0(t)$.) The constraint on the value of λ ensures that t_+^{λ} is a function which is locally integrable on $\mathbb{R}$. Hence, in terms of the conceptual framework developed in Chapter 4, we can say that for any value of $\lambda > 0$ there exists a time-invariant linear system, which we shall represent by the operator I^λ, whose impulse response function is $t_+^{\lambda-1}/\Gamma(\lambda)$. For any continuous input signal $x(t)$ which vanishes outside a finite interval, this T.I.L.S. has a well defined response $y(t)$ given by a classical convolution integral:

$$x(t) \quad \to \quad y(t) \equiv I^\lambda x(t) = \frac{1}{\Gamma(\lambda)} \int_{-\infty}^{+\infty} x(\tau) t_+^{\lambda-1}(t-\tau) d\tau$$

$$= \frac{1}{\Gamma(\lambda)} \int_{-\infty}^{t} x(\tau)(t-\tau)^{\lambda-1} d\tau. \tag{7.10}$$

Moreover a direct computation from the Laplace transform integral gives

$$\mathcal{L}[t_+^{\lambda-1}(t)] = \int_0^{+\infty} t^{\lambda-1} e^{-st} dt = \int_0^{+\infty} \left(\frac{\tau}{s}\right)^{\lambda-1} e^{-\tau} \left(\frac{1}{s}\right) d\tau$$

$$= \frac{1}{s^\lambda} \int_0^{+\infty} \tau^{\lambda-1} e^{-\tau} d\tau = \frac{\Gamma(\lambda)}{s^\lambda}, \tag{7.11}$$

so that the system I^λ has transfer function $\frac{1}{s^\lambda}$. It follows at once that when two such systems I^{λ_1} and I^{λ_2} are connected in cascade then we have

$$x(t) \quad \to \quad y(t) = I^{\lambda_1}\{I^{\lambda_2}x(t)\} = I^{\lambda_1+\lambda_2}x(t). \tag{7.12}$$

When $\lambda = n$, a positive integer, the T.I.L.S. represented by I^n has impulse response function $t_+^{(n-1)}/(n-1)!$ and transfer function $1/s^n$, and carries out an n-fold integration on any admissible input function.

$$x(t) \quad \rightarrow \quad y(t) \equiv I^n x(t)$$

$$= \frac{1}{(n-1)!} \int_{-\infty}^{+\infty} x(\tau)u(t-\tau)(t-\tau)^{n-1}d\tau,$$

(See Chapter 4, Exercises II, No.3). However, the description of I^λ as an integration operator continues to make sense for all positive values of λ, even when λ is not an integer. If, for example, we take $\lambda = 1/2$ then the cascade connection of two systems of the form $I^{1/2}$ is equivalent to a single ideal integrator (see Chapter 4, Exercises II, No:4), and it is therefore natural to refer to $I^{1/2}$ as a 'half-integrator'.

For $\lambda = 0$ the definition (7.10) fails. Setting $\lambda = 0$ in $u_0(t)t^{\lambda-1}/\Gamma(\lambda)$ gives $u_0(t)/t$. This is not a locally integrable function and the convolution integral of (7.10) does not exist in general. On the other hand, as λ tends to zero, the Laplace transform of (7.11) tends to a perfectly well defined limit

$$\lim_{\lambda \to 0} \left[\frac{\Gamma(\lambda)}{s^\lambda}\right] = \frac{\Gamma(0)}{s^0} = 1.$$

Hence we can define I^0 as the operator representing the T.I.L.S. whose transfer function is the constant function 1, and which therefore transmits every admissible input unchanged:

$$x(t) \rightarrow y(t) \equiv I^0 x(t) = \int_{-\infty}^{+\infty} x(\tau)\delta(t-\tau)d\tau = x(t). \qquad (7.13)$$

The role of impulse response function is now played by the delta function, and the classical convolution integral of (7.10) is replaced by the symbolic one of (7.13).

7.1.4 Fractional differentiation.

We have achieved a satisfactory definition of an integration operator I^λ which makes sense for all non-negative values of λ. The next step is to introduce a corresponding definition of a derivative operator of non-integer order. This will effectively extend the meaning of I^λ to negative values of λ. In what follows we shall assume that all input signals $x(t)$ are sufficiently well behaved to allow the necessary integrations and differentiations to be carried out; that is we assume not only that each $x(t)$ vanishes outside some finite interval but also that it may be differentiated as often as we wish.

To begin with, let α be any real number such that $0 \leq \alpha < 1$. Then $I^{1-\alpha}$ is certainly well-defined and so we can define an operator D^α which represents the cascade connection of the fractional integrator represented by $I^{1-\alpha}$ and an ideal differentiator:

$$x(t) \quad \to \quad y(t) \equiv D^\alpha x(t) = \frac{d}{dt} I^{1-\alpha} x(t)$$

$$= \frac{1}{\Gamma(1-\alpha)} \frac{d}{dt} \int_{-\infty}^{t} x(\tau)(t-\tau)^{-\alpha} d\tau. \tag{7.14}$$

For $\alpha = 0$ this reduces to

$$x(t) \quad \to \quad D^0 x(t) = \frac{d}{dt} \int_{-\infty}^{t} x(\tau) d\tau = x(t)$$

so that the impulse response is $\delta(t)$ and D^0 is identical with I^0. Further, since we have the equivalent representation

$$D^\alpha x(t) = \frac{1}{\Gamma(1-\alpha)} \int_{0}^{+\infty} \tau^{-\alpha} x'(t-\tau) d\tau = I^{1-\alpha} x'(t),$$

it follows at once that

$$I^\alpha \{ D^\alpha x(t) \} = I^\alpha \{ I^{1-\alpha} x'(t) \} = I^1 x'(t) = x(t).$$

Hence, for $0 \leq \alpha < 1$, D^α is the inverse of the fractional integration operator I^α. Finally, for an arbitrary $\lambda > 0$ we can always write

$$\lambda = n - 1 + \alpha \quad \text{where} \quad n \in \mathbb{N} \quad \text{and} \quad 0 \leq \alpha < 1,$$

and to complete the definition, we need only to set

$$D^\lambda x(t) \equiv \frac{1}{\Gamma(1-\alpha)} \frac{d^n}{dt^n} \int_{-\infty}^{t} x(\tau)(t-\tau)^{-\alpha} d\tau. \tag{7.15}$$

The operator D^λ represents a (causal) time-invariant linear system which consists of a cascade combination of a fractional integrator of order $1-\alpha$ and n ideal differentiators. It should therefore have the transfer function $H(s)=s^{n-1+\alpha}$. However, for any value of $n > 1$, there will exist no (ordinary) function $h(t)$ of which $H(s)$ is the Laplace transform, and so we cannot express (7.11) in the form of a classical convolution integral. If $\alpha = 0$ then D^λ becomes D^{n-1} and the problem is solved at least formally by the introduction of the generalised function $\delta^{(n-1)}(t)$. To obtain a satisfactory development of a comprehensive **fractional calculus** we therefore need to define other types of generalised functions.

7.2 HADAMARD FINITE PART

7.2.1 To begin with we recall once more the elementary formula for integration by parts. If f and g are standard functions on $\mathbb{R}$ which are continuously differentiable everywhere, and if $[a,b]$ is any finite, closed, real interval then

$$\int_a^b f(t)g'(t)dt = f(b)g(b) - f(a)g(a) - \int_a^b f'(t)g(t)dt.$$

In particular if f happens to vanish identically outside $[a,b]$ then this becomes

$$\int_{-\infty}^{+\infty} f(t)g'(t)dt = [f(t)g(t)]_a^b - \int_a^b f'(t)g(t)dt$$

$$= \int_{-\infty}^{+\infty} [-f'(t)]g(t)dt. \tag{7.16}$$

Equation (7.16) remains true if we relax the conditions on g and stipulate merely that it be absolutely continuous on $[a,b]$: that is, that there exists a function $h(t)$ which is integrable over $[a,b]$ and such that

$$g(t) - g(a) = \int_a^t h(\tau)d\tau$$

for all $t \in [a,b]$. (See Problem 6 of Exercises I of Chapter 2.) In this case g need not be differentiable everywhere. As an example take the function

$$g(t) = t_+^{1/2}(t) \equiv u_0(t)\sqrt{t}.$$

This is an absolutely continuous function on $(-\infty, +\infty)$ since we have

$$u_0(t)\sqrt{t} = \int_{-\infty}^t \frac{u_0(\tau)}{2\sqrt{\tau}}d\tau$$

for all t. It is differentiable everywhere except at the origin, and its classical derivative, $t_+^{-1/2}(t)/2 \equiv u_0(t)/2\sqrt{t}$, is actually unbounded on any neighbourhood of the origin. Nevertheless, as is easily confirmed, equation (7.16) holds for any continuously differentiable function f which vanishes identically outside some finite interval. (The integral on the left-hand side of (7.16) always exists because the derivative $g'(t)$ is a locally integrable function.)

7.2.2 If instead we now take $g(t) = u(t)$ then only the integral on the right-hand side of equation (7.16) exists in the classical sense, and we have

$$\int_{-\infty}^{+\infty} [-f'(t)]g(t)dt = \int_0^b [-f'(t)]dt = [-f(t)]_0^b = f(0).$$

This value is, of course, precisely what we would wish to attribute to the symbolic integral which, in the formal delta function calculus, now does duty for the left-hand side of equation (7.16). In fact we could equally well have based the definition of the delta function itself on the equation

$$\int_{-\infty}^{+\infty} f(t)\delta(t)dt = \int_{-\infty}^{+\infty} [-f'(t)]u(t)dt \qquad (7.17)$$

in the sense that the meaning of the left-hand side is to be understood simply as the value which the (well-defined) right-hand side assigns to it. The only difference from what has gone before is that the operation of the delta function would have been limited by (7.17) to a more restricted class of functions f than we previously found desirable.

7.2.3 The introduction of delta functions allows the processes and operations of elementary calculus to be extended to functions which may have simple, jump discontinuities. Now let us use a similar technique to extend the concept of derivative so that it applies usefully and meaningfully to functions with infinite discontinuities. As a first example we shall take the function

$$g(t) = -2t_+^{-1/2}(t) \equiv -2u_0(t)/\sqrt{t}.$$

This is differentiable in the classical sense everywhere except at the origin, and its classical derivative is given, for all $t \neq 0$, by

$$g'(t) = u_0(t)/t^{3/2}.$$

The function $g'(t)$ is unbounded in any neighbourhood of the origin. In this respect the situation appears at first to be similar to that of the absolutely continuous function $u_0(t)\sqrt{t}$, and its derivative $u_0(t)/2\sqrt{t}$, already discussed in Sec. 7.2.1. However, this time the derivative $g'(t)$ is not a locally integrable function: it is not integrable over any finite interval containing the origin. Hence if we write

$$\int_{-\infty}^{+\infty} f(t)g'(t)dt = \int_{-\infty}^{+\infty} [-f'(t)][-2u_0(t)/\sqrt{t}]dt \qquad (7.18)$$

then this will not be a valid equation for all continuously differentiable functions f which vanish outside a finite interval. The right-hand side always exists, but the left-hand side generally does not. More precisely, for the right-hand side we have

$$\int_{-\infty}^{+\infty} [-f'(t)][-2u_0(t)/\sqrt{t}]dt = 2\int_0^b \frac{f'(t)}{\sqrt{t}}dt \qquad (7.19)$$

for some finite number b, and this is an absolutely convergent integral. On the other hand we can analyse the left-hand side of (7.18) as follows:

$$\int_{-\infty}^{+\infty} f(t)[u_0(t)t^{-3/2}]dt = \lim_{\varepsilon\downarrow 0} \int_\varepsilon^{+\infty} f(t)t^{-3/2}dt$$

$$= -2\lim_{\varepsilon\downarrow 0}\int_\varepsilon^b f(t)\frac{d}{dt}\left(\frac{1}{\sqrt{t}}\right)dt = -2\lim_{\varepsilon\downarrow 0}\left[\frac{-f(\varepsilon)}{\sqrt{\varepsilon}} - \int_\varepsilon^b \frac{f'(t)}{\sqrt{t}}dt\right] \qquad (7.20)$$

making an entirely legitimate use of integration by parts and noting that $f(b) = 0$.

Now, since f is continuously differentiable, we can appeal to the First Mean Value Theorem of the differential calculus (that is Taylor's Theorem with $n = 1$) and write

$$f(t) = f(0) + tf'(\theta t) \qquad (7.21)$$

where θ is some number lying between 0 and 1. In particular this gives

$$\frac{f(\varepsilon)}{\sqrt{\varepsilon}} = \frac{f(0)}{\sqrt{\epsilon}} + \sqrt{\varepsilon}f'(\theta\varepsilon).$$

It follows that $f(\varepsilon)/\sqrt{\varepsilon}$ tends to 0 with ϵ if and only if $f(0) = 0$. If $f(0) \neq 0$ then this ratio tends either to $+\infty$ or to $-\infty$. Hence if $f(0) = 0$ the limit of (7.20) exists and is equal to the finite value given by (7.19); in this case the equation (7.18) is a valid one.

If, on the other hand, $f(0) \neq 0$ then the form of (7.20) allows us to separate out the required 'finite part' of the now divergent integral which appears as the left-hand side of (7.18). To do this we have only to remove the term $2f(\varepsilon)/\sqrt{\varepsilon}$ which, as the above argument shows, is what causes the limit in (7.20) to become infinite. The remaining well-defined quantity is called the **Hadamard Finite Part** of the divergent integral concerned, and we write

$$Fp \int_{-\infty}^{+\infty} f(t)[u_0(t)t^{-3/2}]dt$$

$$\equiv \lim_{\varepsilon \downarrow 0} \left[\int_{\varepsilon}^{+\infty} f(t)t^{-3/2}dt - 2f(\varepsilon)/\sqrt{\varepsilon} \right], \qquad (7.22)$$

the symbol 'Fp' expressing the fact that we are extracting a certain finite number from an integral which is actually divergent.

7.2.4 The preceding discussion may be summarised as follows. We require to define an 'operational' derivative, $Dg(t)$, of the function $g(t) = -2u_0(t)/\sqrt{t}$. This must satisfy the equation

$$\int_{-\infty}^{+\infty} f(t)[Dg(t)]dt = \int_{-\infty}^{+\infty} [-f'(t)]g(t)dt$$

at least for all continuously differentiable functions f which vanish outside some finite interval. We cannot identify $Dg(t)$ with the ordinary derivative $g'(t)$ since this generally makes the integral on the left-hand side divergent. Instead we agree that $Dg(t)$ shall denote a generalised function of a new type. In older texts it is usual to refer to $Dg(t)$ as a **pseudo-function** and to denote it by the symbol $\mathrm{Pf}\{u(t)t^{-3/2}\}$. Alternatively we can now extend the definition of the expression $t_+^\lambda(t)$ introduced in section 7.1.3 and write $Dg(t)$ as $t_+^{-3/2}(t)$. This pseudo-function $\mathrm{Pf}\{u(t)t^{-3/2}\} \equiv t_+^{-3/2}(t)$ is characterised by the operation which carries each continuously differentiable function f, vanishing outside some finite interval, into the number given by the limit in (7.22). It needs to be carefully distinguished from the ordinary function $u(t)t^{-3/2}$ and the notation $t_+^{-3/2}$ offers a convenient way of doing so.

We can express the finite part of the integral in (7.22) in a more convenient form by using again the Taylor expansion of (7.21). We have

$$\int_{-\infty}^{+\infty} f(t)[u_0(t)t^{-3/2}]dt = \lim_{\varepsilon \downarrow 0} \int_{\varepsilon}^{+\infty} f(t)t^{-3/2}dt$$

$$= \lim_{\varepsilon \downarrow 0} \left[\int_{\varepsilon}^{+\infty} \frac{f(0)}{t^{3/2}}dt + \int_{\varepsilon}^{+\infty} \frac{f'(\theta t)}{\sqrt{t}}dt \right]$$

$$= \lim_{\varepsilon \downarrow 0} \left[\frac{2f(0)}{\sqrt{\varepsilon}} + \int_{\varepsilon}^{+\infty} \frac{f'(\theta t)}{\sqrt{t}}dt \right]. \qquad (7.23)$$

Comparing (7.23) with (7.22) it is clear that we could use any one of the following formulations:

$$\int_{-\infty}^{+\infty} f(t)\mathrm{Pf}\{u_0(t)/t^{3/2}\}dt \equiv \mathrm{Fp}\int_{-\infty}^{+\infty} f(t)\{u_0(t)/t^{3/2}\}dt$$

$$= \int_0^{+\infty} \{f'(\theta t)/\sqrt{t}\}dt = \int_0^{+\infty} \frac{f(t)-f(0)}{t^{3/2}}dt. \qquad (7.24)$$

7.3　PSEUDO-FUNCTIONS

7.3.1 The analysis leading to the definition of the pseudo-function $\mathrm{Pf}\{u_0(t)t^{-3/2}\}$ can easily be generalised. Let $h(t) = u_0(t)/t^{\alpha+1}$ where $0 < \alpha < 1$. Take an arbitrary continuously differentiable function f, vanishing outside some finite interval, say $[a, b]$. Then

$$\int_\varepsilon^{+\infty} \frac{f(t)}{t^{\alpha+1}}dt = \int_\varepsilon^{+\infty} \left[\frac{f(0)}{t^{\alpha+1}} + \frac{f'(\theta t)}{t^\alpha}\right] dt$$

$$= \int_\varepsilon^{+\infty} \frac{f'(\theta t)}{t^\alpha}dt + f(0)\left[\frac{t^{-\alpha}}{-\alpha}\right]_\varepsilon^{+\infty} = \int_\varepsilon^{+\infty} \frac{f'(\theta t)}{t^\alpha}dt + \frac{f(0)}{\alpha\varepsilon^\alpha} \qquad (7.25)$$

where we make use of the Taylor expansion

$$f(t) = f(0) + tf'(\theta t), \quad 0 < \theta < 1.$$

The last integral on the right-hand side of (7.25) converges absolutely as ε tends to zero, and the second term tends to plus or minus infinity unless $f(0) = 0$.
Hence

$$\mathrm{Fp}\int_{-\infty}^{+\infty} f(t)h(t)dt = \lim_{\varepsilon\downarrow 0}\int_\varepsilon^{+\infty} \frac{f'(\theta t)}{t^\alpha}dt$$

$$= \lim_{\varepsilon\downarrow 0}\int_\varepsilon^{+\infty}\left[\frac{f(t)-f(0)}{t^{\alpha+1}}\right]dt. \qquad (7.26)$$

This integral, which is absolutely convergent, defines the pseudo-function $\mathrm{Pf}\{u_0(t)/t^{\alpha+1}\}$, and we may write this as $t_+^{-1-\alpha}$ in order to distinguish it from the ordinary function $u(t)/t^{\alpha+1}$. This pseudo-function is the generalised derivative of the ordinary function $-u(t)/\alpha t^\alpha$, as can be seen from the following analysis.

We have

$$\int_\varepsilon^{+\infty} \left[\frac{f(t) - f(0)}{t^{\alpha+1}} \right] dt =$$

$$\left[\left(\frac{t^{-\alpha}}{-\alpha} \right) (f(t) - f(0)) \right]_\varepsilon^{+\infty} + \frac{1}{\alpha} \int_\varepsilon^{+\infty} f'(t) t^{-\alpha} dt$$

which reduces to

$$\int_\varepsilon^{+\infty} [-f'(t)] \left[\frac{t^{-\alpha}}{-\alpha} \right] dt + \frac{1}{\alpha} \left[\frac{f(\varepsilon) - f(0)}{\varepsilon^\alpha} \right]$$

where the last term tends to 0 with ε since $f(\varepsilon) - f(0) = \varepsilon f'(\theta \varepsilon)$ and $0 < \alpha < 1$. As a result we may write

$$\int_{-\infty}^{+\infty} f(t) \mathrm{Pf} \left\{ \frac{u_0(t)}{t^{\alpha+1}} \right\} dt \equiv \mathrm{Fp} \int_{-\infty}^{+\infty} \left[f(t) \frac{u_0(t)}{t^{\alpha+1}} \right] dt$$

$$= \int_{-\infty}^{+\infty} [-f'(t)] \left[\frac{-u_0(t)}{\alpha t^\alpha} \right] dt \qquad (7.27)$$

7.3.2 We can apply a similar analysis to the case of a function $g(t) = u_0(t)/t^{\alpha+2}$, where $0 < \alpha < 1$. As before we resort to a Taylor expansion in order to separate out the required finite part of a normally divergent integral. For an arbitrary twice-continuously differentiable function f (again assumed to vanish outside some finite interval) we have

$$f(t) = f(0) + t f'(0) + \frac{t^2}{2} f''(\theta t)$$

where θ lies between 0 and 1. Then we get

$$\int_\varepsilon^{+\infty} \frac{f(t)}{t^{\alpha+2}} dt = \int_\varepsilon^{+\infty} \left[\frac{f(0)}{t^{\alpha+2}} + \frac{f'(0)}{t^{\alpha+1}} + \frac{f''(\theta t)}{2t^\alpha} \right] dt$$

$$= \int_\varepsilon^{+\infty} \frac{f''(\theta t)}{2t^\alpha} dt + \frac{f'(0)}{\alpha \varepsilon^\alpha} + \frac{f(0)}{(\alpha+1)\varepsilon^{\alpha+1}}$$

which diverges as ε tends to 0 unless $f'(0) = f(0) = 0$.

This allows us to write the Hadamard finite part of the divergent integral $\int_{-\infty}^{+\infty} f(t) g(t) dt$ in the following form:

$$\mathrm{Fp} \int_{-\infty}^{+\infty} f(t)g(t)dt = \lim_{\varepsilon \downarrow 0} \int_{\varepsilon}^{+\infty} \frac{f''(\theta t)}{2t^\alpha}dt$$

$$= \int_0^{+\infty} \left[\frac{f(t) - f(0) - tf'(0)}{t^{\alpha+2}} \right] dt. \qquad (7.28)$$

The mapping which carries each given f into the number specified by (7.28) defines the pseudo-function $\mathrm{Pf}\{u_0(t)/t^{\alpha+2}\}$, which we may write as $t_+^{-2-\alpha}$. Also, integrating by parts,

$$\int_{\varepsilon}^{+\infty} \left[\frac{f(t) - f(0) - tf'(0)}{t^{\alpha+2}} \right] dt$$

$$= \left[\frac{t^{-(\alpha+1)}}{-(\alpha+1)} \{f(t) - f(0) - tf'(0)\} \right]_{\varepsilon}^{+\infty}$$

$$+ \frac{1}{\alpha+1} \int_{\varepsilon}^{+\infty} \left[\frac{f'(t) - f'(0)}{t^{\alpha+1}} \right] dt$$

$$= \frac{1}{\alpha+1} \int_{\varepsilon}^{+\infty} \left[\frac{f'(t) - f'(0)}{t^{\alpha+1}} \right] dt$$

$$+ \frac{1}{\alpha+1} \left[\frac{f(\varepsilon) - f(0) - \varepsilon f'(0)}{\varepsilon^{\alpha+1}} \right].$$

In the limit, as ε tends to zero, the last integral is seen, from (7.26), to converge to the limit

$$\int_0^{+\infty} \left[\frac{f'(t) - f'(0)}{(\alpha+1)t^{\alpha+1}} \right] dt$$

$$= \int_{-\infty}^{+\infty} [-f'(t) + f'(0)] \left[\frac{-u_0(t)}{(\alpha+1)t^{\alpha+1}} \right] dt$$

$$\equiv \int_{-\infty}^{+\infty} [-f'(t)] \mathrm{Pf} \left\{ \frac{-u_0(t)}{(\alpha+1)t^{\alpha+1}} \right\} dt.$$

The remaining terms tend to zero with ε since

$$f(\varepsilon) - f(0) - \varepsilon f'(0) = \frac{\varepsilon^2}{2} f''(\theta \varepsilon)$$

and $0 < \alpha < 1$. Thus $\mathrm{Pf}\{u_0(t)/t^{\alpha+2}\}$ behaves as the generalised derivative of the pseudo-function $\mathrm{Pf}\{-u_0(t)/(\alpha+1)t^{\alpha+1}\}$ in the sense that

$$\int_{-\infty}^{+\infty} f(t)\mathrm{Pf}\left\{ \frac{u_0(t)}{t^{\alpha+2}} \right\} dt = \int_{-\infty}^{+\infty} [-f'(t)]\mathrm{Pf}\left\{ \frac{-u_0(t)}{(\alpha+1)t^{\alpha+1}} \right\} dt$$

at least for every twice-continuously differentiable function f which vanishes outside some finite interval. Note that it is now generally necessary to compute Hadamard finite parts of divergent integrals in order to obtain both sides of the 'integration by parts' equation.

7.3.3 Clearly we could go on to develop similar pseudo-functions of the form $t_+^{-n-\alpha} \equiv \mathrm{Pf}\{u_0(t)/t^{\alpha+n}\}$, where n is an integer and $0 < \alpha < 1$. Then the action of $\mathrm{Pf}\{u_0(t)/t^{\alpha+n}\}$ on an arbitrary n-times continuously differentiable function $f(t)$ would be given by

$$\int_{-\infty}^{+\infty} f(t)\mathrm{Pf}\left\{\frac{u_0(t)}{t^{\alpha+n}}\right\}dt$$

$$\equiv \int_0^{+\infty} t^{-n-\alpha}\left\{f(t) - \sum_{k=0}^{n-1}\frac{f^{(k)}(0)}{k!}t^k\right\}dt \qquad (7.29)$$

This allows us to express the whole range of (real) integration operators I^λ in terms of a (generalised) convolution operation. Thus, for $\lambda > 0$ we have

$$I^\lambda x(t) = \frac{1}{\Gamma(\lambda)}\int_{-\infty}^{+\infty} x(t-\tau)t_+^{\lambda-1}(\tau)d\tau$$

$$= \frac{1}{\Gamma(\lambda)}\int_0^{+\infty} x(t-\tau)\tau^{\lambda-1}d\tau$$

where $t_+^{\lambda-1}$ is an ordinary locally integrable function and the convolution integral is to be taken in the classical sense.

For $\lambda = 1 - \alpha - n$, where $0 < \alpha < 1$ and n is a non-negative integer, $t_+^{\lambda-1}$ now denotes a pseudo-function and we have

$$I^\lambda x(t) = \frac{1}{\Gamma(\lambda)}\int_{-\infty}^{+\infty} x(t-\tau)t_+^{-n-\alpha}(\tau)d\tau$$

$$= \frac{1}{\Gamma(\lambda)}\int_0^{+\infty} \tau^{-n-\alpha}\left\{x(t-\tau) - \sum_{k=0}^{n-1}\frac{(-1)^k x^{(k)}(t)}{k!}\tau^k\right\}d\tau.$$

Finally, for $\lambda = 0, -1, -2, \ldots$ we have the now familiar conventional representation

$$I^{-n}x(t) = \int_{-\infty}^{+\infty} x(t-\tau)\delta^{(n)}(\tau)d\tau.$$

7.3.4 For completeness we consider briefly the case of the pseudo-functions corresponding to ordinary functions of the form $u(t)/t^m$ where m is a positive integer. The same general principles apply, but there are points of detail which demand special attention. We consider first the function $u(t)/t$. For all $t \neq 0$ this is the classical derivative of the (locally integrable) function $u_0(t) \log(|t|)$, but it is not itself a locally integrable function. If we write

$$\int_{\varepsilon}^{+\infty} \frac{f(t)}{t} dt = \int_{1}^{+\infty} \frac{f(t)}{t} dt + \int_{\varepsilon}^{1} \left[\frac{f(0)}{t} + f'(\theta t) \right] dt$$

$$= \int_{1}^{+\infty} \frac{f(t)}{t} dt + \int_{\varepsilon}^{1} f'(\theta t) dt - f(0) \log(\varepsilon),$$

then the required Hadamard finite part is unequivocally defined as

$$\mathrm{Fp} \int_{-\infty}^{+\infty} f(t) \frac{u_0(t)}{t} dt$$

$$= \lim_{\varepsilon \downarrow 0} \left[\int_{1}^{+\infty} \frac{f(t)}{t} dt + \int_{\varepsilon}^{1} \left\{ \frac{f(t) - f(0)}{t} \right\} dt \right] \qquad (7.30)$$

An integration by parts gives

$$\mathrm{Fp} \int_{-\infty}^{+\infty} f(t) \frac{u_0(t)}{t} dt$$

$$= \lim_{\varepsilon \downarrow 0} \left[\int_{\varepsilon}^{+\infty} \{-f'(t)\} \log(t) dt - \{f(\epsilon) - f(0)\} \log(\varepsilon) \right]$$

and we have only to note that $f(\varepsilon) - f(0) = f'(\theta \varepsilon)$ where $|f'(\theta \varepsilon)|$ is bounded and $\lim_{\varepsilon \downarrow 0} (\varepsilon \log(\varepsilon)) = 0$. Thus the locally integrable function $u(t) \log(|t|)$ has for its operational derivative the pseudo-function $\mathrm{Pf}\{u(t)/t\}$:

$$\int_{-\infty}^{+\infty} f(t) \mathrm{Pf} \left\{ \frac{u_0(t)}{t} \right\} dt \equiv \mathrm{Fp} \int_{-\infty}^{+\infty} f(t) \frac{u_0(t)}{t} dt$$

and we can legitimately write

$$\int_{-\infty}^{+\infty} f(t) D\{u(t) \log(|t|)\} dt = \int_{-\infty}^{+\infty} \{-f'(t)\} u_0(t) \log(|t|) dt. \qquad (7.31)$$

Similarly the operational or generalised derivative of the locally integrable function $u(-t)\log(|t|)$ is the pseudo-function $\text{Pf}\{u(-t)/t\}$. Further, since

$$\log(|t|) = u(t)\log(|t|) + u(-t)\log(|t|)$$

we should have

$$D\{\log(|t|)\} = \text{Pf}\left\{\frac{u(t)}{t}\right\} + \text{Pf}\left\{\frac{u(-t)}{t}\right\} \equiv \text{P}\left\{\frac{1}{t}\right\},$$

where

$$\int_{-\infty}^{+\infty} f(t)\text{P}\left(\frac{1}{t}\right) dt = \int_0^1 \frac{f(t) - f(0)}{t} dt + \int_1^{+\infty} \frac{f(t)}{t} dt$$

$$+ \int_0^1 \frac{f(0) - f(-t)}{t} dt - \int_1^{+\infty} \frac{f(-t)}{t} dt$$

$$= \int_0^{+\infty} \frac{f(t) - f(-t)}{t} dt. \tag{7.32}$$

The final (absolutely convergent) integral in (7.32) can be expressed in the alternative form,

$$\lim_{\varepsilon \downarrow 0} \int_\varepsilon^{+\infty} \frac{f(t) - f(-t)}{t} dt$$

$$= \lim_{\varepsilon \downarrow 0} \left\{ \int_\varepsilon^{+\infty} \frac{f(t)}{t} dt + \int_{-\infty}^{-\varepsilon} \frac{f(t)}{t} dt \right\},$$

which is just the Cauchy Principal Value of $\int_{-\infty}^{+\infty} \frac{f(t)}{t} dt$.

7.3.5 For $m = 2, 3, \ldots$, the position is complicated by the somewhat unexpected appearance of delta functions. If, for example, we take the case $m = 2$ then proceeding as above we get

$$\int_\varepsilon^{+\infty} \frac{f(t)}{t^2} dt = \int_1^{+\infty} \frac{f(t)}{t^2} dt + \int_\varepsilon^1 \left[\frac{f(0)}{t^2} + \frac{f'(0)}{t} + \frac{f''(\theta t)}{2} \right] dt$$

$$= \int_1^{+\infty} \frac{f(t)}{t^2} dt + \int_\varepsilon^1 \frac{f''(\theta t)}{2} dt - f(0) + \left[\frac{f(0)}{\varepsilon} - f'(0)\log(\varepsilon) \right]$$

so that

$$\text{Fp} \int_{-\infty}^{+\infty} f(t) \frac{u_0(t)}{t^2} dt$$

$$= \lim_{\varepsilon \downarrow 0} \left[\int_1^{+\infty} \frac{f(t)}{t^2} dt - f(0) + \int_\varepsilon^1 \frac{f''(\theta t)}{2} dt \right]$$

$$= \int_1^{+\infty} \frac{f(t)}{t^2} dt - f(0) + \lim_{\varepsilon \downarrow 0} \int_\varepsilon^1 \frac{f(t) - f(0) - t f'(0)}{t^2} dt$$

which reduces on integration by parts to

$$\int_1^{+\infty} \frac{f'(t)}{t} dt + \lim_{\varepsilon \downarrow 0} \left[\int_\varepsilon^1 \frac{f'(t) - f'(0)}{t} dt + \frac{f(\varepsilon) - f(0)}{\varepsilon} \right].$$

Using (7.27) it follows that

$$\int_{-\infty}^{+\infty} f(t) \text{Pf} \left\{ \frac{u_0(t)}{t^2} \right\} dt$$

$$= \int_1^{+\infty} \frac{f'(t)}{t} dt + \int_0^1 \frac{f'(t) - f'(0)}{t} dt + f'(0)$$

$$= - \int_{-\infty}^{+\infty} \{-f'(t)\} \text{Pf} \left\{ \frac{u(t)}{t} \right\} dt - \int_{-\infty}^{+\infty} f(t) \delta'(t) dt.$$

That is,

$$\text{Pf} \left\{ \frac{u_0(t)}{t^2} \right\} = -\frac{d}{dt} \text{Pf} \left\{ \frac{u(t)}{t} \right\} - \delta'(t).$$

In general it can be shown that

$$\frac{d}{dt} \text{Pf} \left\{ \frac{u_0(t)}{t^m} \right\} = -\text{Pf} \left\{ \frac{m u(t)}{t^{m+1}} \right\} + \frac{(-1)^m}{m!} \delta^{(m)}(t). \qquad (7.32)$$

Further reading.

Most books on distributions devote some space to pseudo-functions of the type discussed in this chapter, often as an illustration of the existence of generalised functions which are not delta functions. The book by Sir James Lighthill referred to in Chapter 2 contains an extensive discussion with applications to asymptotic estimation of Fourier transforms. For further information on fractional calculus the treatise **Fractional Integrals and Derivatives** by **Samko**, **Kilbas** and **Marichev** (Gordon and Breach, 1996) is recommended.

Chapter 8

Introduction to distributions

8.1 TEST FUNCTIONS

8.1.1 A systematic and comprehensive theory of generalised functions which includes pseudo-functions such as those discussed in the preceding chapter, as well as delta functions and derivatives of delta functions, was developed by **Laurent Schwartz** in the 1950s. In its original form, as presented in Schwartz's **Theorie des Distributions**, the theory is not easy, and some familiarity with topological vector spaces and other such recondite material from functional analysis is necessary. Various alternative treatments of distributions have since been developed but the subject remains a difficult one for the newcomer. We give here a simplified outline of the essential features of the Schwartz theory, which is necessarily very incomplete and which is intended primarily as a preparation for the study of more advanced texts. Other approaches to distributions will be reviewed subsequently and in particular a nonstandard interpretation of the theory is offered in ths final chapter of this book.

To begin with we need to specify a basic class of functions for which the characteristic operations of all the generalised functions under consideration are always well defined. We shall refer to the members of this basic class as **test functions**.

So far as the Dirac delta function is concerned the essential requirement for a test function is that it should be continuous. For an nth order derivative of a delta function we need test functions which are continuosly differentiable at least up to order n. Hence, in order to deal

with derivatives of the delta function of arbitrary order, the basic class
of test functions should contain only functions which are infinitely differentiable. But even this is not enough. The operations which characterise
the unit step function, and the pseudo-functions t_+^λ introduced in Chapter 7, are ultimately defined in terms of integrals which should converge.
To meet this condition we restrict attention to test functions which each
vanish outside some finite interval. Accordingly we are led to consider
a fundamental set, $\mathcal{D}$, of ordinary functions which enjoy the following
properties

(i) each function f in $\mathcal{D}$ is infinitely differentiable,

(ii) for each function f in $\mathcal{D}$ we can find a corresponding finite, closed
interval $[a, b]$ outside which f vanishes identically: f is then said to have
its **support** contained in $[a, b]$.

Such functions do exist, and an important family of members of $\mathcal{D}$ can
be defined as follows: first, let

$$\psi(t) = \begin{cases} \exp[(t^2 - 1)^{-1}] & \text{for } |t| < 1 \\ 0 & \text{for } |t| \geq 1 \end{cases} \tag{8.1}$$

and set $A \equiv \int_{-1}^{+1} \psi(t)dt$. Then, for $n = 1, 2, \ldots$, write $\psi_n(t) \equiv \frac{n}{A}\psi(nt)$,
so that

$$\psi_n(t) = \begin{cases} \frac{n}{A} \exp\left(\frac{1}{n^2 t^2 - 1}\right), & \text{for } |t| < 1/n \\ 0 & \text{for } |t| \geq 1/n \end{cases} \tag{8.2}$$

It is easily confirmed that each function ψ_n vanishes identically outside
the finite interval $[-1/n, +1/n]$, has continuous derivatives of all orders
and, incidentally, is such that $\int_{-\infty}^{+\infty} \psi_n(t)dt = 1$.

8.1.2 If f and g are members of $\mathcal{D}$ then so also is the function h defined
by pointwise addition,

$$h(t) = f(t) + g(t).$$

Similarly if f belongs to $\mathcal{D}$ and if α is any real number then αf is also a
member of $\mathcal{D}$, where αf denotes the function whose value at each point
t is $\alpha f(t)$.

Thus $\mathcal{D}$ is closed under addition and under multiplication by real numbers, and therefore has an algebraic structure analogous to that displayed
by the three-dimensional space of ordinary experience when we treat it
as a space of **vectors**. Given any two points in ordinary space, say

$x = (x_1, x_2, x_3)$ and $y = (y_1, y_2, y_3)$, we define their **vector sum** $x + y$ to be the point (or vector) specified by

$$z \equiv x + y = (x_1 + y_1, x_2 + y_2, x_3 + y_3).$$

Again, if α is any real number then we define the product αx to be the point

$$\alpha x = (\alpha x_1, \alpha x_2, \alpha x_3).$$

It is easily confirmed that these operations of **vector addition** and **multiplication by scalars** will have the following properties:

L1. Vector addition is associative, commutative and distributive with respect to multiplication by scalars,

$$x + (y + z) = (x + y) + z \ ; \quad x + y = y + x \ ;$$

$$\alpha(x + y) = \alpha x + \alpha y.$$

L2. For any vector x and any scalars α and β we have

$$(\alpha + \beta)x = \alpha x + \beta x \quad ; \quad \alpha(\beta x) = (\alpha \beta)x.$$

L3. There is a unique vector 0, called the **null vector**, which is such that

$$x + 0 = 0 + x = x \ \text{ for every vector } x.$$

L4. To each vector x there corresponds a unique vector $(-x)$ which is such that

$$x + (-x) = 0.$$

L5. For every vector x it is the case that

$$1x = x \quad \text{and} \quad 0x = 0.$$

In general any set of objects which is closed under an operation (called vector addition) and under a form of multiplication by real, or complex, numbers such that conditions L1-L5 are satisfied, is called a **linear space** and its members are called **vectors**. When the scalar multipliers are confined to the real number system the space is called a **real linear space**; otherwise it is called a **complex linear space**. Clearly the set of test functions $\mathcal{D}$ with addition and multiplication by real numbers defined as above does constitute a real linear space in this sense. We could always obtain a corresponding complex linear space by considering instead the set of all complex-valued functions f where

$$f(t) = f_1(t) + if_2(t)$$

and f_1 and f_2 are functions belonging to the real linear space $\mathcal{D}$.

8.1.3 The analogy between $\mathcal{D}$ and ordinary three-dimensional space can be taken a stage further. If $x = (x_1, x_2, x_3)$ then we would normally understand the length of the vector x to be the number

$$\|x\| = +\sqrt{\{x_1^2 + x_2^2 + x_3^2\}}.$$

Conceptually this is just the same as the 'distance from the origin' of the point whose coordinates are x_1, x_2 and x_3. We can go on to speak of the distance between two arbitrary points, say (x_1, x_2, x_3) and (y_1, y_2, y_3) as the length of the vector $x - y$:

$$d(x, y) \equiv \|x - y\| = +\sqrt{\{(x_1 - y_1)^2 + (x_2 - y_2)^2 + (x_3 - y_3)^2\}}.$$

The concept of distance allows us to talk meaningfully about convergence and limiting processes in space in general, since there will be a specific sense in which we can speak of a sequence of points (or vectors) approaching more and more closely to some limiting point.

Similar ideas of distance and convergence can be developed for the space $\mathcal{D}$ of infinitely differentiable functions. However to do this it is necessary to isolate those essential formal properties which characterise 'length' and 'distance'. We begin with the definition of a **norm**.

A norm is a real-valued function $\|x\|$, defined on a linear space $\mathcal{E}$, which satisfies the following conditions:

N1. $\|x\| \geq 0$ for every x in $\mathcal{E}$, and $\|x\| = 0$ if and only if x is the null vector of $\mathcal{E}$.

N2. $\|x + y\| \leq \|x\| + \|y\|$, (the 'triangle inequality'),

N3. $\|\alpha x\| = |\alpha| \|x\|$, for every vector x and any scalar α.

A linear space $\mathcal{E}$, together with a norm $\|x\|$ defined on $\mathcal{E}$, is usually referred to as a **normed linear space**. The distance, $d(x, y)$, between any two members x and y is then defined as

$$d(x, y) = \|x - y\|$$

and the real-valued function $d(x, y)$ is spoken of as a **metric** on $\mathcal{E}$. A metric can be defined from first principles as a real-valued function satisfying the conditions

D1. $d(x, y) \geq 0$ for every x in $\mathcal{E}$, and $d(x, y) = 0$ if and only if $x = y$.

D2. $d(x, y) = d(y, x)$.

D3. $d(x, z) \leq d(x, y) + d(y, z)$.

Any set of objects on which such a function $d(x, y)$ is defined is said to be a **metric space**. Clearly every linear space with a norm is automatically a metric space with the metric defined in terms of the norm as above. Once a norm has been defined on a linear space there is at once a corresponding sense of convergence in that space. In a metric space in general a sequence of points $(x_m)_{m \in \mathbf{N}}$ is said to converge, in the sense of the metric, to the point x as its limit if, given any real number $\varepsilon > 0$, we have

$$d(x, x_m) < \varepsilon$$

for all but finitely many values of m. That is to say, x is the limit of the x_m if and only if the distance between x and the x_m tends to zero as m tends to infinity:

$$\lim_{m \to \infty} x_m = x \quad \text{if and only if} \quad \lim_{m \to \infty} d(x, x_m) = 0.$$

When, as here, the metric is defined in terms of a norm, the criterion for convergence becomes

$$\lim_{m \to \infty} \|x - x_m\| = 0.$$

Convergence in this mode is often referred to as 'convergence in norm'.

8.1.4 There are many ways in which we might think of defining a norm, and hence a specific mode of convergence, on the linear space $\mathcal{D}$. To begin with, consider the so-called **supremum norm** or **uniform norm**,

$$\|f\| = \sup_t |f(t)|. \tag{8.3}$$

The fact that (8.3) does define a norm is easy to show. In the first place it is clear that $\|f\| \geq 0$ for every f in $\mathcal{D}$ and that $\|f\| = 0$ if and only if $f(t) = 0$ for all t. Also

$$\|\alpha f\| = \sup_t |\alpha f(t)| = |\alpha| \sup_t |f(t)| = |\alpha| \|f\|$$

which establishes N3. Finally, the triangle inequality N2 follows since

$$|f(t) + g(t)| \le |f(t)| + |g(t)| \quad \text{for each } t,$$

so that

$$\sup\{|f(t) + g(t)|\} \le \sup\{|f(t)| + |g(t)|\}$$
$$\le \sup|f(t)| + \sup|g(t)|.$$

Now suppose that $(f_m)_{m \in \mathbb{N}}$ is any sequence of functions in $\mathcal{D}$ which converges to a limit f in the sense of the norm (7.32). Then

$$\lim_{m \to \infty} \|f - f_m\| = \lim_{m \to \infty} \{\sup|f(t) - f_m(t)|\} = 0$$

and the convergence is seen to be necessarily uniform. Note however that this does not mean that the limit function f is necessarily itself a member of $\mathcal{D}$. The uniformity of the convergence certainly ensures that f is a bounded continuous function but not that it is differentiable. We cannot even conclude that f has support contained in some finite interval, but only that $|f(t)|$ tends to zero as $|t|$ becomes arbitrarily large:

$$\lim_{|t| \to \infty} |f(t)| = 0.$$

On the other hand if each of the functions f_m vanishes identically outside the same fixed finite interval $[a, b]$ then it is obvious that the limit function f must also vanish outside $[a, b]$. To distinguish this special case we shall refer to the convergence of the f_m as **locally restricted uniform convergence**.

8.1.5 Instead of the uniform norm on $\mathcal{D}$ we might choose to define a norm $\|f\|^{(1)}$ as follows:

$$\|f\|^{(1)} = \max\{\|f\|, \|f'\|\} = \max\{\sup_t|f(t)|, \sup_t|f'(t)|\}. \qquad (8.4)$$

Convergence in the sense of this norm is stronger than ordinary uniform convergence. For if a sequence (f_m) converges to a limit f in the sense of (8.4) then we must have both

$$\lim_{m \to \infty} \{\sup|f(t) - f_m(t)|\} = 0$$

and

$$\lim_{m \to \infty} \{\sup|f'(t) - f'_m(t)|\} = 0.$$

That is, (f_m) converges uniformly to f and, in addition, the sequence (f'_m) of the first derivatives of the f_m converges uniformly to the limit f'. Once again convergence in this sense does not necessarily ensure that the limit f is a member of $\mathcal{D}$. In general we can only conclude that f must always be at least continuously differentiable and that both $|f(t)|$ and $|f'(t)|$ tend to zero as $|t|$ tends to ∞.

More generally, for any given positive integer p, we can define a norm $\|f\|^{(p)}$ on $\mathcal{D}$ by writing

$$\|f\|^{(p)} = \max\{\|f\|, \|f'\|, \ldots, \|f^{(p)}\|\}$$

$$= \max_{0 \leq k \leq p}\{\sup|f^{(k)}(t)|\} \tag{8.5}$$

The mode of covergence appropriate to this norm is most conveniently referred to as p-**uniform convergence**. A sequence $(f_m)_{m\in\mathbb{N}}$ is said to converge p-uniformly to the limit function f if and only if, for $k = 0, 1, \ldots, p$, each of the sequences $(f_m^{(k)})_{m\in\mathbb{N}}$ of the kth derivatives of the f_m converges uniformly to the corresponding derivative $f^{(k)}$ of the limit function f. We shall describe this as **locally restricted p-uniform convergence** if, in addition, each of the functions f_m vanishes identically outside the same fixed interval $[a, b]$.

8.2 FUNCTIONALS AND DISTRIBU-TIONS

Armed with these possible norms and corresponding modes of convergence on the linear space $\mathcal{D}$, we can now consider how to characterise all the generalised functions encountered so far. Each of them has been associated with a certain mapping, defined for an appropriate class of ordinary functions and taking numerical values. In this section we shall be more specific about the nature and properties of such mappings, particularly in relation to the algebraic structure of the basic space of test functions $\mathcal{D}$ and to the various modes of convergence which can be defined within it.

8.2.1 Given a linear space $\mathcal{E}$ let μ denote a mapping which assigns to each member f of $\mathcal{E}$ a certain well-defined number which we will write as $\mu(f)$, or sometimes as $< \mu, f >$. Any such numerically valued function on $\mathcal{E}$ is called a **functional**. We say that μ is a **linear functional** on E if

$$\mu(\alpha f + \beta g) = \alpha\mu(f) + \beta\mu(g) \tag{8.6}$$

for every f and g in $\mathcal{E}$ and every pair of scalars (numbers) α and β. If, in addition, $\mathcal{E}$ is a normed linear space and there exists some fixed, finite, positive number K such that

$$|\mu(f)| \leq K\|f\| \tag{8.7}$$

for every f in $\mathcal{E}$, then μ is said to be a **bounded** linear functional on $\mathcal{E}$. Every bounded linear functional on $\mathcal{E}$ is necessarily **continuous** in the sense that whenever a sequence (f_m) in $\mathcal{E}$ converges in norm to a limit f in $\mathcal{E}$ then the corresponding numerical sequence $(\mu(f_m))$ converges in the usual sense to the number $\mu(f)$. This follows at once from the fact that

$$|\mu(f_m) - \mu(f)| = |\mu(f - f_m)| \leq K\|f - f_m\|,$$

since, by hypothesis, $\|f - f_m\|$ tends to 0 as m tends to infinity. The set of all bounded linear functionals on a given space $\mathcal{E}$ turns out to be a normed linear space in its own right, under the following definitions of vector addition, scalar multiplication and norm:

Addition. The sum $(\mu_1 + \mu_2)$ of two bounded linear functionals μ_1 and μ_2 is the bounded linear functional μ defined by

$$\mu(f) = \mu_1(f) + \mu_2(f), \quad \text{for every } f \in \mathcal{E}.$$

Multiplication by scalars. The product $\alpha\mu$, where μ is a bounded linear functional on $\mathcal{E}$ and α is a scalar, is defined by

$$(\alpha\mu)(f) = \alpha\mu(f), \quad \text{for every } f \in \mathcal{E}.$$

Norm. The norm of a bounded linear functional μ is defined to be

$$\|\mu\| = \sup_{\|f\| \leq 1} |\mu(f)|. \tag{8.8}$$

It is easy to see that the above conditions of addition and scalar multiplication do satisfy the required conditions for the operations of a linear space, and it is only a little more troublesome to confirm that $\|\mu\|$ also satisfies the conditions for a norm.

In what follows we shall be primarily concerned with real linear functionals defined on real linear spaces. However, it is a trivial exercise to extend the results to the case of complex-valued linear functionals defined on real (or complex) linear spaces.

8.2.2 Consider first the linear space $\mathcal{D}$ with the metric structure imposed by the norm

$$\|f\| = \sup_t |f(t)|.$$

Let μ be a linear functional defined on $\mathcal{D}$ which is bounded with respect to this norm. That is to say we assume that $\mu(f)$ is well-defined for every function f belonging to $\mathcal{D}$ and that there exists some finite number K such that

$$|\mu(f)| \le K\|f\| = K\sup_t |f(t)| \tag{8.9}$$

for all f in $\mathcal{D}$.

Examples.

(i) Let h be a fixed function which is absolutely integrable over the whole range $-\infty < t < +\infty$. That is, h is such that

$$\int_{-\infty}^{+\infty} |h(t)|\,dt < +\infty.$$

Then the mapping

$$f \longrightarrow \int_{-\infty}^{+\infty} f(t)h(t)\,dt$$

is certainly well-defined for all f in $\mathcal{D}$ and is plainly linear. Moreover we have

$$\left| \int_{-\infty}^{+\infty} f(t)h(t)\,dt \right| \le \int_{-\infty}^{+\infty} |f(t)||h(t)|\,dt$$

$$\le \sup_t |f(t)| \int_{-\infty}^{+\infty} |h(t)|\,dt$$

so that the mapping is bounded with respect to the uniform norm. If we write

$$\mu_h(f) \equiv <\mu_h, f> = \int_{-\infty}^{+\infty} f(t)h(t)\,dt$$

then it follows that μ_h is a linear functional on $\mathcal{D}$, bounded with respect to the uniform norm, and that

$$\|\mu_h\| = \int_{-\infty}^{+\infty} |h(t)|\,dt. \tag{8.10}$$

(ii) For any fixed real number a the mapping $f \longrightarrow f(a)$ obviously defines a linear functional on $\mathcal{D}$ which is bounded with respect to the uniform norm. We shall now use the symbol δ_a to denote this functional and refer to it as the **Dirac delta functional located at** $t = a$:

$$\delta_a(f) \equiv < \delta_a, f >= f(a) \text{ with } \|\delta_a\| = 1. \tag{8.11}$$

The functional δ_a carries out the sampling operation assigned in the elementary naive treatment to the translate $\delta(t-a)$ of the delta function. Where convenient, however, we shall still follow the usual practice of accepting the (purely symbolic) integral notation

$$\delta_a(f) \equiv \int_{-\infty}^{+\infty} f(t)\delta(t-a)dt = f(a).$$

(iii) Finite linear combinations of delta functions and absolutely integrable functions can be represented in the same way in terms of linear functionals on $\mathcal{D}$, bounded with respect to the uniform norm. However we need an extension of these ideas if we are to obtain like representations of the generalised functions defined by the unit step function $u(t)$ (which is not absolutely integrable over the whole range $-\infty < t < +\infty$) or of infinite linear combinations of delta functions such as

$$\delta(\sin(t)) \equiv \sum_{m=-\infty}^{+\infty} \delta(t - m\pi).$$

Consider again the linear functional defined by the mapping

$$f \longrightarrow \int_{-\infty}^{+\infty} f(t)h(t)dt$$

where h is now not necessarily absolutely integrable over the whole interval $(-\infty, +\infty)$ but is merely locally integrable; that is, such that

$$\int_a^b |h(t)|dt < +\infty$$

for any finite interval $[a, b]$. Given any f in $\mathcal{D}$ there must exist some finite interval $[a, b]$ outside which f vanishes identically. Hence

$$\left| \int_{-\infty}^{+\infty} f(t)h(t)dt \right| = \left| \int_a^b f(t)h(t)dt \right| \leq \|f\| \int_a^b |h(t)|dt.$$

In general we say that a linear functional is **relatively** or **locally** bounded with respect to the uniform norm on $\mathcal{D}$ if, for each finite real interval $I = [a, b]$, we can find a finite positive number $K = K(I)$ such that

$$|\mu(f)| \le K(I) \sup_t |f(t)| \qquad (8.12)$$

for every function f in $\mathcal{D}$ with support contained in I. In the above example we have

$$K(I) = \int_a^b |h(t)| \, dt.$$

Again, if

$$\mu = \delta(\sin(t)) = \sum_{m=-\infty}^{+\infty} \delta(t - m\pi)$$

then for any finite interval $[a, b]$ there will exist integers n_1, n_2, such that

$$(n_1 - 1)\pi \le a < b \le (n_2 + 1)\pi.$$

Hence for any f in $\mathcal{D}$ with support contained in $I = [a, b]$ we have

$$|\mu(f)| = \left| \sum_{m=n_1}^{n_2} f(m\pi) \right| \le \sum_{m=n_1}^{n_2} |f(m\pi)| \le (n_2 - n_1 + 1)\|f\|$$

so that μ is relatively bounded with respect to the uniform norm.

Note that boundedness always implies relative boundedness but not conversely. A relatively bounded linear functional on $\mathcal{D}$ will actually be bounded on $\mathcal{D}$ only if the numbers $K(I)$ in (8.12) are themselves uniformly bounded: that is, only if there exists some finite number $K(\mathbb{R})$ such that $K(I) \le K(\mathbb{R})$ for all finite intervals I contained in $\mathbb{R}$.

8.2.3 Let μ be any linear functional on $\mathcal{D}$ which is relatively bounded for the uniform norm. We denote by $D\mu$ the linear functional on $\mathcal{D}$ which is given by

$$D\mu(f) = \mu(-f') \qquad (8.13)$$

for all f in $\mathcal{D}$. $D\mu$ will be referred to as the **derivative** of the functional μ. If, in particular, μ is of the form μ_h where h is a locally integrable function which has a locally integrable classical derivative h', then an integration by parts shows that for any f in $\mathcal{D}$,

$$\int_{-\infty}^{+\infty} f(t) h'(t) \, dt = \int_{-\infty}^{+\infty} \{-f'(t)\} h(t) \, dt.$$

That is, this notion of derivative is compatible with the classical definition used in elementary calculus whenever the latter exists.

In general, if f is any function in $\mathcal{D}$ which vanishes outside an interval $I = [a, b]$, then

$$|D\mu(f)| = |\mu(-f')| \leq K(I)\sup_t|f'(t)| \leq K(I)\|f\|^{(1)}.$$

Thus $D\mu$ will certainly always be relatively bounded with respect to the norm $\|f\|^{(1)}$, though not necessarily with respect to the uniform norm $\|f\|$ itself.

Examples.

(i) The derivative of δ_a is defined by the mapping

$$f \longrightarrow \delta_a(-f') = f'(a).$$

Then, trivially,

$$|D\delta_a(f)| = |f'(a)| \leq \|f'\| \leq \|f\|^{(1)}.$$

On the other hand, given any positive number M, however large, we can always find some f in $\mathcal{D}$ such that $|f(t)| \leq 1$ for all t but with $|f'(a)| > M$. This is enough to show that $D\delta_a \equiv \delta_a'$ is not relatively bounded for the uniform norm.

(ii) The locally integrable function $u(t)\log(|t|)$ has the (generalised) derivative $\text{Pf}\{u(t)/t\}$. For any function f in $\mathcal{D}$ it is the case that

$$\int_{-\infty}^{+\infty} f(t)\text{Pf}\{u(t)/t\}dt \equiv \text{Fp} \int_{-\infty}^{+\infty} f(t)\frac{u(t)}{t}dt$$

$$= \int_{-\infty}^{+\infty} \{-f'(t)\}u(t)\log(|t|)dt$$

and it follows that the functional defined by the pseudo-function $\text{Pf}\{u(t)/t\}$ is the derivative, in the sense of (8.13), of the functional defined by the locally integrable function $u(t)\log(|t|)$. What is more we have

$$\left| \int_{-\infty}^{+\infty} f(t)\text{Pf}\{u(t)/t\}dt \right| = \left| \int_0^{+\infty} \frac{f(t) - f(0)}{t}dt \right|$$

$$= \left| \int_0^b f'(\theta t)dt \right| \leq b.\sup_t|f'(t)|$$

for some b and some θ such that $0 < \theta < 1$. Hence the functional concerned is relatively bounded with respect to the norm $\|f\|^{(1)}$.

8.2.4 By applying the differentiation process defined by (8.13) to linear functionals relatively bounded with respect to the norm $\|f\|^{(1)}$ we would similarly generate linear functionals relatively bounded with respect to the norm $\|f\|^{(2)}$ and so on. This leads us to suggest a tentative definition of a generalised function as:

any linear functional on $\mathcal{D}$ which is relatively bounded with respect to some norm $\|f\|^{(p)}$, where $p = 0, 1, 2, \ldots$.

Following Schwartz we call such a functional a **distribution**: strictly, we should call such a functional a **distribution of finite order**, because the definition used by Schwartz is more general and includes linear functionals other than those discussed so far. [see § 8.4, below]

8.3   **CALCULUS OF DISTRIBUTIONS**

8.3.1 Any distribution defined by a locally integrable function h is said to be a **regular** distribution. In many contexts we may identify the function h with the distribution μ_h which it defines, though, as shown below, this must be done with due caution. Any other kind of distribution is said to be **singular**. From what has been said about functionals in general it is a straightforward matter to summarise the algebra of distributions.

(i) **Equality.** Two distributions μ_1 and μ_2 are said to be equal if and only if $\mu_1(f)$ and $\mu_2(f)$ are equal for every $f \in \mathcal{D}$. If μ_{h_1} and μ_{h_2} are regular distributions defined by locally integrable functions h_1 and h_2 respectively then $\mu_1 = \mu_2$ does not necessarily imply that $h_1(t) = h_2(t)$ for all t. The most we can say is that

$$\int_a^b |h_1(t) - h_2(t)| dt = 0$$

for every finite interval $[a, b]$, and therefore that the functions are equal almost everywhere, $h_1(t) =_{a.e} h_2(t)$.

(ii) The **null distribution** 0 is the linear functional which maps every function f in $\mathcal{D}$ into zero:

$$0(f) = 0, \quad \text{for every } f \in \mathcal{D}.$$

It is a regular distribution which can be defined by, or represented by, any function h which vanishes almost everywhere: that is, any function h such that

$$\int_a^b |h(t)|dt = 0$$

for every finite interval $[a, b]$.

A distribution μ is said to be zero on any open interval in $\mathbb{R}$ if $\mu(f) = 0$ for every function $f \in \mathcal{D}$ with support contained in that interval. Now the definition of support for a (continuous) function can be conveniently re-phrased as follows:

a function $h(t)$ has support contained in an interval $[a, b]$ if $h(t) = 0$ for all t in $(-\infty, a)$ and for all t in $(b, +\infty)$

This allows to give a definition of support for a distribution μ, whether it is regular or singular:

Support of a distribution. A distribution μ is said to have its support contained in an interval $[a, b]$ if $\mu(f) = 0$ for every function $f \in \mathcal{D}$ with support contained in $(-\infty, a)$ and for every function $f \in \mathcal{D}$ with support contained in $(b, +\infty)$. If $[a, b]$ is finite then μ is said to be a distribution of bounded (or compact) support.

The set of all distributions of bounded support is an important and useful subset. In particular note that the delta function $\delta(t - a)$ located at the point $t = a$ is certaimly a distribution of bounded support, and that its support is clearly just the point $t = a$ itself.

(iii) **Addition** and **multiplication by scalars** are defined for distributions as for linear functionals in general. That is to say, we define

$$(\mu_1 + \mu_2)(f) \equiv \mu_1(f) + \mu_2(f)$$

$$(\alpha\mu)(f) \equiv \alpha\{\mu(f)\}$$

for all f in $\mathcal{D}$, where μ_1, μ_2 and μ are distributions and α is any scalar. In particular note that for any distribution μ we get $\mu + 0 = 0 + \mu = \mu$, and $0(\mu) = 0$.

(iv) **Multipliers.** A function ϕ is called a **multiplier** for a distribution μ if $\mu(\phi f)$ is well defined (as a continuous linear functional) for every function f for which $\mu(f)$ exists. If ϕ is a multiplier for μ then the product $\phi\mu$ is defined to be the distribution given by setting $(\phi\mu)(f) \equiv \mu(\phi f)$ for every $f \in \mathcal{D}$. For a general distribution μ the product $\phi\mu$ is well defined for any function ϕ in $\mathcal{D}$. For particular distributions the product may be defined for a wider range of functions. For example, as we saw in Chapter 3,

$$(\phi\delta)(f) = \phi(0)f(0) \equiv \phi(0)\delta(f)$$

for any continuous function ϕ. Similarly, if ϕ is continuously differentiable, then the product $\phi\delta'$ is well defined by

$$(\phi\delta')(f) \equiv \delta'(\phi f) = -\phi'(0)f(0) - \phi(0)f'(0)$$

$$\equiv -\phi'(0)\delta(f) + \phi(0)\delta'(f).$$

In the case of regular distributions there is an obvious close correspondence between the algebraic operations carried out on the distributions themselves and those carried out on the locally integrable functions which define those distributions. Now as already remarked we cannot strictly identify a regular distribution μ with a specific function h. At best we can only identify μ with a certain equivalence class $[f]$ of locally integrable functions, any two members h_1, h_2 of which will be equal almost everywhere. The regular distribution associated with the unit step function, for example, could equally well be defined by any one of the locally integrable functions $u_c(t)$, but should not be identified with any one of them.

The importance of distinguishing between the functions and processes of ordinary calculus and the generalised functions, or distributions, and the corresponding operations on them is most clearly shown by the problems encountered in trying to define products of distributions. Unless restrictions are placed on the factors involved the associative law may fail. For example, the product $t\delta(t)$ is plainly equal to the null distribution, 0, and so we should have

$$\frac{1}{t}(t\delta(t)) = 0.$$

On the other hand it seems equally clear that

$$\left(\frac{1}{t}t\right)\delta(t) = 1\delta(t) = \delta(t).$$

8.3.2 Convergence. A sequence (μ_n) of distributions is said to converge to the distribution μ as its limit if and only if

$$\mu(f) = \lim_{n \to \infty} \mu_n(f)$$

for every f in $\mathcal{D}$. The most familiar examples of convergence in this sense occur when the delta function is represented in terms of a sequence of ordinary functions, as in Chapter 2 above. More generally it can be shown that the action of any distribution μ can be similarly represented. That is, for any distribution μ we can find a sequence $(h_n)_{n \in \mathbb{N}}$ of ordinary functions such that, for every test function f in $\mathcal{D}$ we have

$$\mu(f) = \lim_{n \to \infty} \int_{-\infty}^{+\infty} h_n(t) f(t) dt,$$

and μ may be identified with an equivalence class of such sequences. This can be made the basis of an alternative *definition* of distributions.

Finally note that the infinite series of distributions $\sum_{n=1}^{+\infty} \mu_n$ converges to the distribution μ as its sum if and only if the sequence of partial sums

$$\left(\sum_{k=1}^{n} \mu_k \right)$$

converges to μ as its limit. That is to say,

$$\mu = \sum_{n=1}^{+\infty} \mu_n \quad \text{if and only if} \quad \mu = \lim_{n \to \infty} \sum_{k=1}^{n} \mu_k.$$

The notation $\sum_{n=-\infty}^{+\infty} \delta(t - nT)$ used throughout this book for an infinite periodic train of delta functions (or, rather, delta distributions) is in conformity with this definition.

8.4   GENERAL SCHWARTZ DISTRIBUTIONS

8.4.1 Distributions of finite order. The theory of generalised functions outlined in this chapter makes use only of the more accessible parts of modern analysis associated with the elementary theory of normed linear spaces. As a result we have only been able to define a restricted sub-class of the generalised functions introduced by Laurent Schwartz and called by him distributions. We have only discussed what are properly described as **distributions of finite order**. Thus, to begin with, we dealt with regular distributions (those defined by ordinary functions and integration processes) and delta functions: these were all identified as linear functionals which are relatively bounded on $\mathcal{D}$ with respect to the uniform norm $\|f\|$, and they are accordingly referred to as **distributions of order** 0. Next we have distributions like the first derivative δ' of the delta function, which are linear functionals relatively bounded on $\mathcal{D}$ with respect to the norm $\|f\|^{(1)}$, and not with respect to the uniform norm $\|f\|$ itself: these are similarly described as **distributions of order** 1. And so on, with distributions of order p, where $p \in \mathbb{N}$, being defined as just those linear functionals which are relatively bounded on $\mathcal{D}$ with respect to the norm $\|f\|^{(p)}$, and not with respect to any norm $\|f\|^{(q)}$ where $q < p$.

This restriction is not of any immediate major importance: distributions of finite order are generally the most widely applicable and in practice will be found to suffice for most practical purposes. But for completeness we need to give here some account of the full scope of the Schwartz theory, as well as of its limitations and of some of the subsequent modifications and extensions of the theory which have been proposed.

8.4.2 Schwartz distributions. Consider first the problem of defining general Schwartz distributions - that is, to define the class of all continuous linear functionals on $\mathcal{D}$, including "distributions of infinite order". Briefly, suppose we take the norm $\|f\|^{(p)}$ defined by equation (8.5), and the corresponding concept of p-convergence, and then allow p to tend to infinity. On the one hand we cannot define a norm $\|f\|^{(\infty)}$ which is in any sense a 'limit' of the norms $\|f\|^{(p)}$. On the other, however, the concept of p-uniform convergence can be extended in a meaningful way:

A sequence $(f_n)_{n \in \mathbb{N}}$ in $\mathcal{D}$ may be said to converge Ω-**uniformly** to the limit f if and only if each of the sequences $(f_n^{(k)})_{n \in \mathbb{N}}$ of the kth derivatives of the f_n (for $k = 0, 1, 2, \ldots$) converge uniformly to the kth derivative of f. If all the f_n vanish outside the same fixed interval $[a, b]$ then the limit function f also vanishes outside $[a, b]$ and the convergence is said, as before, to be locally restricted.

With this sense of convergence the linear space $\mathcal{D}$ has a structure which is akin to, but more general than, that defined by a norm or a metric. A linear functional μ on $\mathcal{D}$ is now said to be a **distribution in the sense of Schwartz** if it is continuous under locally restricted Ω-uniform convergence on $\mathcal{D}$; that is, if we have

$$\mu(f) = \lim_{n \to \infty} \mu(f_n)$$

for every sequence (f_n) of functions in $\mathcal{D}$ such that

(i) for $k = 0, 1, 2, \ldots$, each sequence $(f_n^{(k)})_{n \in \mathbb{N}}$ converges uniformly to the limit $f^{(k)}$, and

(ii) there exists some finite interval $[a, b]$ outside which each of the functions f_n vanishes identically.

Plainly, if μ is any linear functional which is relatively bounded on $\mathcal{D}$ with respect to some norm $\|f\|^{(p)}$, where $p = 0, 1, 2, \ldots$. then μ is certainly continuous under locally restricted Ω-convergence on $\mathcal{D}$. That is to say, all the generalised functions which we have encountered so far, and which we have now agreed to call distributions of finite order, are distributions in the sense of Schwartz. We denote by $\mathcal{D}'$ the vector space of all Schwartz distributions, and by $\mathcal{D}'_{fin}$ the vector subspace of all distributions of finite order.

8.4.3 Distributions and ultradistributions. In Chapter 6 we were able to derive some ad hoc extensions of the classical Fourier transform which applied to the unit step function, delta functions, end even to infinite series of delta functions. It is clearly desirable that there should be a canonical definition of the Fourier Transform, consistent with classical definitions, which is applicable to all distributions - or, at least to some large and well-defined subset of distributions. Now for the classical Fourier transform of sufficiently well-behaved functions we have

$$\tilde{f}(t) = \int_{-\infty}^{+\infty} f(\tau)e^{-it\tau}d\tau; \quad f(\tau) = \int_{-\infty}^{+\infty} \tilde{f}(t)e^{it\tau}dt$$

where, for convenience, we now use the well-known alternative notation $\tilde{f}$ for the transform of the function f. Further, for such well behaved functions f and g the Parseval relation holds:

$$\int_{-\infty}^{+\infty} f(t)\tilde{g}(t)dt = \int_{-\infty}^{+\infty} \tilde{f}(t)g(t)dt.$$

To extend the definition of Fourier transform to distributions Schwartz used a generalised form of the Parseval relation to define the transform of a distribution μ as the functional $\tilde{\mu}$ satisfying

$$< \tilde{\mu}, \phi >=< \mu, \tilde{\phi} > .$$

But there is a difficulty here: if $\phi \in \mathcal{D}$ then its well-defined classical Fourier Transform $\tilde{\phi}$ will certainly be an infinitely differentiable function, but it will not be a function which vanishes outside some finite interval. That is to say, $\tilde{\phi}$ will not be a member of the space $\mathcal{D}$. It belongs to another, quite different function space $\mathcal{Z} \equiv \mathcal{Z}(\mathbb{R})$. Dually, if $\tilde{\phi}$ does belong to $\mathcal{D}$ then its inverse Fourier transform ϕ will be a member of $\mathcal{Z}$. It can be shown, in fact, that $\mathcal{Z}$ and $\mathcal{D}$ have no members in common: $\mathcal{D} \cap \mathcal{Z}$ is the empty set.

It follows that if μ is an arbitrary Schwartz distribution then we can certainly define its Fourier transform $\tilde{\mu}$ as a linear continuous functional on the space $\mathcal{Z}$, but it will not necessarily be defined as a linear continuous functional on $\mathcal{D}$. It may therefore not be a Schwartz distribution at all. The linear continuous functionals on $\mathcal{Z}$ in fact constitute a new space of generalised functions called **ultradistributions**. Some ultradistributions do turn out to be distributions as well, and an extension of the Fourier transform can be successfully defined for them - as the examples discussed in Chapter 6 illustrate. These constitute an important subspace of $\mathcal{D}'$ and are called **tempered distributions**. But there are ultradistributions which are not distributions and there also exist distributions which are not ultradistributions. Further discussion of this matter is outside the scope of this text, but a very clear account of the distributional transform and of the definition and properties of ultradistributions is to be found in the book by Zemanian referred to below.

8.4.4 Distributions and Sato Hyperfunctions. Another approach to the problem of generalising the Fourier Transform stems from the work of **Carlmann** in the 1940s, and a very brief sketch of the essential ideas can be given as follows. (We need to assume some acquaintance with complex variable theory here.) Carlmann observed that if a function f, not necessarily absolutely integrable over $\mathbb{R}$, satisfies a condition of the form

$$\int_0^x |f(y)|dy = 0(|x|^\kappa)$$

for some natural number κ, and if we write

$$g_1(z) = \int_{-\infty}^{0} f(y)e^{-ixy}dy$$

and

$$g_2(z) = -\int_{0}^{+\infty} f(y)e^{-ixy}dy$$

then $g_1(z)$ is analytic for all $\mathrm{Im}(z) > 0$ and $g_2(z)$ is analytic for all $\mathrm{Im}(z) < 0$. Moreover, for $\beta > 0$, the function

$$g(z) \equiv g_1(x + i\beta) - g(x - i\beta)$$

is the classical Fourier transform of the function $e^{-\beta|t|}f(t)$. The original function f can be recovered by taking the inverse Fourier transform of g and multipying by $e^{\beta|t|}$. This suggested that a route to the generalisation of the Fourier transform could be found by associating with f a pair of functions $f_1(z)$ and $f_2(z)$ analytic in the upper and lower half-planes respectively. This idea forms the basis of the theory of **hyperfunctions** developed by **M.Sato** in 1959/60 (although it was anticipated in some respects by several other mathematicians). In order to give a brief sketch of this theory it is necessary to introduce the following additional notation:

$\mathcal{H}(\mathbb{C}\backslash\mathbb{R})$ = the space of all functions analytic outside the real axis.

$\mathcal{H}(\mathbb{C})$ = subspace of all functions in $\mathcal{H}(\mathbb{C}\backslash\mathbb{R})$ which are entire.

$\mathcal{H}^{p,loc}(\mathbb{C}\backslash\mathbb{R})$ = space of all functions θ in $\mathcal{H}(\mathbb{C}\backslash\mathbb{R})$ which are of arbitrary growth to infinity but locally of polynomial growth to the real axis (that is, such that for each compact subset K of $\mathbb{R}$ there exists $c_K > 0$ and $r_K \in \mathbb{N}_0$ such that

$$|\theta(z)| \leq c_K|\mathrm{Im}(z)|^{r_K}$$

for all $z \in \mathbb{C}$ with $\mathrm{Re}(z) \in K$ and sufficiently small $\mathrm{Im}(z) \neq 0$).

Definition. The **hyperfunctions** of Sato are the members of the quotient space $\mathcal{H}_S(\mathbb{R}) = \mathcal{H}(\mathbb{C}\backslash\mathbb{R})/\mathcal{H}(\mathbb{C})$, that is, the set of all equivalence classes $[\theta]$ where $\theta(z)$ is defined and analytic on $\mathbb{C}\backslash\mathbb{R}$ and θ_1 is equivalent to θ_2 iff $\theta_1 - \theta_2$ is entire.

Sato hyperfunctions are a genuine extension of Schwartz distributions. This is shown by the following crucial result established by **Bremermann** in 1965:

Theorem (Bremermann). If μ is any Schwartz distribution then there exists a function $\mu^0(z)$ defined and analytic in $\mathbb{C}\backslash\mathbb{R}$ such that

$$< \mu, \phi >= \lim_{\epsilon \to 0} \int_{-\infty}^{+\infty} [\mu^0(x+i\epsilon) - \mu^0(x-i\epsilon)]\phi(x)dx.$$

Conversely, if $\theta \in \mathcal{H}^{p,loc}(\mathbb{C}\backslash\mathbb{R})$ then there exists a distribution μ such that $\theta(z) = \mu^0(z)$.

In particular, if μ is a distribution of compact support then $\mu^0(z)$ is given explicitly by

$$\mu^0(z) = \frac{1}{2\pi i} < \mu, (x-z)^{-1} > .$$

For example, $\delta^0(z) = \frac{1}{2\pi i} < \delta, (x-z) >= -\frac{1}{2\pi i z}$, and we have

$$< \delta, \phi >= \lim_{\epsilon \to 0} \frac{\epsilon}{\pi} \int_{-\infty}^{+\infty} \frac{1}{x^2 + \epsilon^2}\phi(x)dx = \phi(0).$$

8.4.5 New Generalised Functions. Some 50 years after its development by Laurent Schwartz the theory of distributions remains one of the most powerful and comprehensive theories of generalised functions available in mathematical analysis. That it does have limitations has already been shown in the preceding discussion of the difficulties presented by a corresponding generalisation of the Fourier transform. But there are other problems with distributions which have been known since the first appearance of the theory, the most notorious and intractable being that of a satisfactory and coherent definition of multiplication for distributions. We have seen that even in the apparently simple case of the product of the delta function and the unit step function, there is what appears to be a fundamental ambiguity: $u(t)\delta(t) = k\delta(t)$, where k is an arbitrary constant. Even the product of step functions seems to cause trouble, for on the one hand we would expect to have the result,

$$Du^2(t) = 2Du(t) = 2\delta(t),$$

whereas it seems clear that $u^2(t) = u(t)$, so that $Du^2(t) = \delta(t)$! Even when distributional products seem to be well defined we have trouble, as shown in Sec.8.3.1 by the failure of associativity in the case of the products,

$$\frac{1}{t}(t\delta(t)) \neq \left(\frac{1}{t}t\right)\delta(t).$$

A great deal of work has been devoted to this multiplication problem over the years but it was not until the 1980s that what can claim to be a genuinely final solution appeared with the development of the so-called **New Generalized Functions** of **Jean Francois Colombeau**. Even a sketch of this theory is beyond the scope of the present text, and despite its manifest importance there has as yet been no really satisfactory account at a really elementary level. (See remarks below.)

Further reading.

Of the specialised texts on generalised functions currently available, the multi-volume work, **Generalized Functions** by **Gel'fand, Shilov** and others, (Academic Press, 1964-66), is still almost certainly the most comprehensive and generally useful. For most practical applications the general reader is likely to find his needs amply satisfied by Chapters I and II of the first volume of the series. It should be made clear, however, that these two chapters actually comprise some 200 pages! The authors themselves claim that these chapters "contain the standard minimum which must be known by all mathematicians and physicists who have to deal with generalized functions."In fact there is considerably more material than the term "standard material"might lead one to expect. The treatment of distributions follows the linear functional approach of Schwartz as outlined above in the present chapter. It is hoped that this very brief sketch may serve as an introduction to the Gel'fand and Shilov text and make the material of their Volume 1 more readily accessible.

Much of the same ground is covered in the book **Distribution Theory and Transform Analysis** by **A.H.Zemanian**, (McGraw-Hill, 1965). This is generally rather easier to follow than Volume 1 of Gel'fand and Shilov and is liberally supplied with exercises, but is almost exclusively concerned (as here) with the case of a single real variable. This is not unreasonable since the applications dealt with by Zemanian are primarily in the field of linear systems and signal processing. This book does contain a good introduction to the theory of ultradistributions.

In the book **the Theory of Generalised Functions** by **D.S.Jones** (Cambridge, 1982) a generalised function is defined as an equivalence class of certain sequences of ordinary functions. As a result the reader is obliged to become familiar with a new terminology (though admittedly a very natural and straightforward one). It is a comprehensive text and is very well supplied with examples, but the general reader is once again likely to find it heavy going to begin with. A most useful and highly readable introduction to the terminology and to the whole general point of view of this sequential approach to generalised functions is to be found in the celebrated short text **An Introduction to Fourier Analysis and Generalised Functions** by **Sir James Lighthill**, of which the fuller treatment given by Jones is a natural developmemt.

Yet another alternative theory of Schwartz distributions was developed by the Portuguese mathematician **Sebastio e Silva**, using an axiomatic approach. (A brief description of the basic idea is given in Chapter 11 of the present text.) Recently an authoritative introductory account of this theory has been published (in English translation) as **Introduction to the Theory of Distributions** by **J.Campos Ferreira**, (Longman, 1997). Although this theory is comparatively little known, it has much to commend it for its simplicity and intrinsic elegance.

The New Generalized Functions of **J.F.Colombeau** are described in his book **Elementary Introduction to New Generalized Functions**, (North-Holland, 1985), but despite the title the subject matter may still present considerable conceptual difficulties. Another approach which may be found easier to follow is proposed in **Theories of Generalised Functions** by **R.F.Hoskins** and **J.Sousa Pinto**, (Horwood Publishing, 2005). This text contains a critical overview of the various approaches to the theory of Schwartz distributions and to other types of generalised function, including ultradistributions and Sato hyperfunctions.

Chapter 9

Integration Theory

9.1 RIEMANN-STIELTJES INTEGRALS

9.1.1 Throughout all the preceding material we have been able to work in terms of the elementary or **Riemann theory** of integration. This theory is adequate for most of the applications of elementary calculus, and indeed even for an introductory account of a relatively advanced topic such as the Schwartz theory of distributions, as given in Chapter 8. But eventually the limitations of the Riemann theory and the need for the more powerful and sophisticated integration theory of **Lebesgue** become apparent. In this chapter we give a brief overview of the essential features of the Lebesgue theory of integration based on the treatment of the integral as a linear functional. This allows us to deal directly with the more general case of the Stieltjes integral at the outset, without the necessity of a preliminary study of measure. Such aspects of measure theory as are likely to be of immediate practical use can then be derived from the existing theory of the integral. In particular the Dirac delta function appears quite naturally in yet another role, this time as the functional generating a measure concentrated at a single point.

9.1.2 The elementary theory of the Riemann-Stieltjes integral of a bounded function f over a finite, closed interval $[a, b]$ was introduced briefly in Chapter 2. We begin here by reviewing it in slightly more detail. Let v be some fixed, monotone increasing function, hereinafter called the **integrator**, and let f be an arbitrary bounded function which we call the **integrand**. As in Chapter 2 we define a partition, P, of $[a, b]$ to be a sub-division of that interval by finitely many points t_k, where $0 \le k \le n$, and

$$a = t_0 < t_1 < t_2 < \ldots < t_n = b.$$

Then for each given partition P of $[a, b]$ we can always form the following sums

$$\text{Lower sum} \qquad s(f, v, P) \equiv \sum_{k=1}^{n} m_k [v(t_k) - v(t_{k-1})]$$

$$\equiv \sum_{k=1}^{n} m_k \Delta_k v$$

$$\text{Upper sum} \qquad S(f, v, P) \equiv \sum_{k=1}^{n} M_k [v(t_k) - v(t_{k-1})]$$

$$\equiv \sum_{k=1}^{n} M_k \Delta_k v$$

where $M_k = \sup\{f(t)\}$ and $m_k = \inf\{f(t)\}$ for $t_{k-1} \le t \le t_k$, and we write

$$\Delta_k v \equiv v(t_k - v_{k-1}).$$

Then we surely have

$$m[v(b) - v(a)] \le s(f, v, P) \le S(f, v, P) \le M[v(b) - v(a)]$$

where $M = \sup\{f(t)\}$ and $m = \inf\{f(t)\}$ for $a \le t \le b$. Thus, as P ranges over all possible partitions of $[a, b]$, the set $\{s\}$ of all lower sums is bounded above and so has a least upper bound. Similarly the set $\{S\}$ of all upper sums is bounded below and so has a greatest lower bound. Hence we can always define the so-called **lower and upper Riemann-Stieltjes integrals** of f with respect to v as follows:

$$\left\{ \int_a^b f(t) dv(t) \right\}_L = \sup\{s(f, v, P)\}$$

$$\left\{ \int_a^b f(t) dv(t) \right\}_U = \inf\{S(f, v, P)\}.$$

The Riemann-Stieltjes lower integral is always less than or equal to the Riemann-Stieltjes upper integral

$$\left\{ \int_a^b f(t) dv(t) \right\}_L \le \left\{ \int_a^b f(t) dv(t) \right\}_U. \qquad (9.1)$$

If these two semi-integrals are equal then the function f is said to be integrable in the Riemann-Stieltjes sense with respect to v over the range $[a, b]$. The Riemann-Stieltjes (RS) integral of f with respect to v over $[a, b]$ is then taken to be the common value of the upper and lower integrals.

9.1.3 A necessary and sufficient condition for the Riemann-Stieltjes integrability of f with respect to v is that, given any $\varepsilon > 0$, we can always find a corresponding partition P_ε for which we have $S - s < \varepsilon$. It follows easily that every function f continuous on $[a, b]$ is necessarily RS-integrable over $[a, b]$ with respect to any monotone increasing integrator v. For, if f is continuous on $[a, b]$ then it is uniformly continuous there, and so we can always find a corresponding positive number $\eta = \eta(\varepsilon)$ such that

$$|f(t) - f(\tau)| < \varepsilon$$

for all points t and τ such that $|t - \tau| < \eta$. Now choose a partition P of $[a, b]$ such that
$$\Delta \equiv \max(t_k - t_{k-1}) < \eta.$$
The corresponding sums, S_ε and s_ε, for such a partition must satisfy

$$S_\varepsilon - s_\varepsilon = \sum_{k=1}^{n}(M_k - m_k)\Delta_k v < \sum_{k=1}^{n}\varepsilon\Delta_k v.$$

But

$$\sum_{k=1}^{n}\Delta_k v = [v(t_1) - v(t_0)] + [v(t_2) - v(t_1)] + \ldots$$

$$+[v(t_n) - v(t_{n-1})] = v(b) - v(a),$$

so that
$$S_\varepsilon - s_\varepsilon < \varepsilon[v(b) - v(a)].$$

That is, $S_\varepsilon - s_\varepsilon$ can be made as small as we wish, and the result follows.

9.1.4 The limitations of the elementary theory can readily be seen from the following example. Using the notation introduced in Chapter 2 for unit step functions, suppose we try to evaluate the Stieltjes integral

$$\int_{-1}^{+1} u_{1/2}(t)\,du_c(t). \tag{9.2}$$

For any partition of $[-1, +1]$ there can be only one of two possibilities

(i) the origin is an interior point of some sub-interval, in which case $S = 1$ and $s = 0$, or

(ii) the origin is a boundary point of two adjacent sub-intervals: this time we have $S = c/2 + (1 - c) = 1 - c/2$, and $s = (1 - c)/2$.

Thus, for every partition of $[-1, +1]$ it will certainly be true that $S - s \geq 1/2$, and so the integral of (9.2) is undefined in the Riemann-Stieltjes sense.

This difficulty stems of course from the fact that integrator and integrand have a common discontinuity at the origin. However there are awkward features of the theory even when the integrator is a smooth function, so that the Stieltjes integral reduces to an ordinary Riemann integral. There is a classical example of the failure of the elementary Riemann theory in the case of a sequence of functions $(f_m)_{m \in \mathbb{N}}$ defined as follows:

$$f_m(t) = \lim_{n \to \infty} [\cos(m!\pi t)]^{2n} \text{ for } 0 \leq t \leq 1. \tag{9.3}$$

Now $|\cos(m!\pi t)| = 1$ if $m!t$ is an integer and is strictly less than 1 otherwise. Hence, allowing n to tend to infinity, we find that $f_m(t) = 1$ for those values of t in $[0, 1]$ for which $m!t$ is an integer and is zero for all other values of t. This means that $f_m(t)$ is non-zero for only finitely many values of t in $[0, 1]$. For any partition P of $[0, 1]$ we will always have a lower sum $s(f_m, t, P) = 0$. Further, the upper sum $S(f_m, t, P)$ can be made as small as we please by choosing any partition with sufficiently small subdivisions. It follows that for each m the function f_m is integrable in the Riemann sense over $[0, 1]$ and that we have

$$\int_0^1 f_m(t)dt = 0.$$

Let $f(t) = \lim_{m \to \infty} f_m(t)$, and now consider the integrability of f in the Riemann sense. First, if t is irrational, then there exists no value of m for which $m!t$ is an integer: hence $f_m(t) = 0$ for every m and so $f(t) = 0$. On the other hand, if $t = p/q$ where p and q are integers, then $m!t$ is an integer for all m such that $m \geq q$, and so $f_m(t) = 1$ for all such m, which means that $f(t) = 1$. Thus $f(t)$, which is sometimes called the **Dirichlet function**, is highly discontinuous, taking the value 1 at every rational point and the value 0 at every irrational point. Since every sub-interval of $[0, 1]$ contains both rational and irrational points it follows that for every partition of $[0,1]$ we must have $s(f,t,P) = 0$ and $S(f,t,P) = 1$. Hence, the integral of f over the interval $[0,1]$ does not exist in the Riemann sense.

It was primarily considerations of this kind which made a more sophisticated theory of integration desirable in the first place. The point is not so much the existence of functions like f which are non-integrable in the Riemann theory, but that such functions can arise as the limits of sequences of functions which are themselves integrable. In what follows we sketch a simple treatment of the Lebesgue approach to integration theory in a form which applies not only to the integrals of ordinary analysis but to Stieltjes integrals as well.

9.2 EXTENSION OF THE ELEMENTARY INTEGRAL

9.2.1 From the elementary definition of the integral it is easy to establish that it enjoys the following properties:

(i) **Linearity.** If f_1 and f_2 are any two bounded functions each integrable over $[a,b]$ with respect to the monotonic increasing integrator v, and if α and β are any real numbers, then the function $f = \alpha f_1 + \beta f_2$ is also integrable with respect to v and

$$\int_a^b f(t)dv(t) = \alpha \int_a^b f_1(t)dv(t) + \beta \int_a^b f_2(t)dv(t). \qquad (9.4)$$

(ii) **Non-negativity.** For any function f, integrable with respect to v over $[a,b]$, which is such that $f(t) \geq 0$ for all t in $[a,b]$ we have

$$\int_a^b f(t)dv(t) \geq 0. \qquad (9.5)$$

(iii) **Absoluteness.** If f is integrable over $[a,b]$ with respect to the monotone increasing function v then so also is the function $|f|$. Moreover we have

$$\left| \int_a^b f(t)dv(t) \right| = \left| \int_a^b f^+(t)dv(t) - \int_a^b f^-(t)dv(t) \right|$$

$$\leq \int_a^b f^+(t)dv(t) + \int_a^b f^-(t)dv(t) = \int_a^b |f(t)|dv(t),$$

where f^+ and f^- denote respectively the positive and negative components of f:

$$f^+(t) = \max\{f(t), 0\} \quad ; \quad f^-(t) = \max\{-f(t), 0\}.$$

(iv) **Boundedness.** If f is integrable over $[a, b]$ with respect to the integrator v, and if $M = \sup|f(t)|$ for $a \leq t \leq b$, then

$$\left|\int_a^b f(t)dv(t)\right| \leq M[v(b) - v(a)]. \tag{9.6}$$

Now let the integrator v be some fixed monotone increasing function and let f be an arbitrary function which is continuous everywhere and which vanishes identically outside some finite interval, say $I = [a, b]$. Then the Stieltjes integral of f with respect to v exists over any interval containing $[a, b]$ and we may actually write

$$\int_a^b f(t)dv(t) = \int_{-\infty}^{+\infty} f(t)dv(t) < +\infty$$

without any ambiguity and without any need to discuss convergence of the (technically improper) integral on the right-hand side.

We can conveniently restate this in the language of Chapter 8. First we can easily show that the set $C_0(I)$ of all continuous functions f which vanish outside the same fixed interval I is a linear space. Next it follows from (9.4) and (9.5) that the mapping

$$f \rightarrow \int_{-\infty}^{+\infty} f(t)dv(t) \tag{9.7}$$

defines a linear functional on the space $C_0(I)$ which is bounded with respect to the uniform norm $\|f\| = \sup_t|f(t)|$. Further, if we denote by C_0 the linear space of all continuous functions f each of which vanishes identically outside some finite real interval (so that C_0 is the union of the spaces $C_0(I)$ as I ranges over all possible finite closed intervals) then (9.7) is seen to define a linear functional, μ, on C_0 which is relatively bounded with respect to the uniform norm. Moreover, since the Schwartz space $\mathcal{D}$ is contained in the space C_0, we may conclude that the linear functional defined by any mapping of the form (8.7) is necessarily a distribution -

more precisely a distribution of order 0. Our object now is to construct an extension of the domain of definition of the Stieltjes integral defined by the integrator v, and we shall do so by making use of the additional information (given in (9.5) above) that the functional is a non-negative one: that is $\mu(f) \geq 0$ for every f such that $f(t) \geq 0$ for all t.

9.2.2 Let μ be any linear functional on C_0, defined by a Stieltjes integral as in (9.7). We shall denote by C_U the set of all functions which are the limits of monotone increasing sequences of functions in C_0. That is to say, a function g belongs to C_U if and only if there exists a sequence $(f_n)_{n\in\mathbf{N}}$ of functions in C_0 such that

(i) $f_n(t) \leq f_{n+1}(t)$, for all t and every n,

(ii) $g(t) = \lim_{n\to\infty} f_n(t)$, for all t.

We can extend μ as a functional on C_U by monotone increasing limits as follows.

If g belongs to C_U let (f_n) be any sequence of functions in C_0 which increases monotonely to g as its limit, and write

$$\mu(g) \equiv \lim_{n\to\infty} \mu(f_n). \tag{9.8}$$

It may be shown that this defines $\mu(g)$ uniquely, no matter which particular sequence (f_n) converging to g we happen to have chosen. Further, it is clear that if g is a function which belongs to C_0 itself then g is necessarily a member of C_U (we need only consider the constant monotone increasing sequence obtained by taking $f_n = g$ for every n). Also the value of $\mu(g)$ given by (9.8) is the same as that originally assigned to it by its definition as a functional on C_0. That is, the extension of μ by means of (9.8) is consistent. In general, of course, the members of C_U will not belong to C_0. Indeed, an arbitrary function g in C_U need not vanish outside a finite interval, need not be continuous everywhere, and may even assume infinite values. In the same way we must accept the possibility that if g belongs to C_U then $\mu(g)$ may turn out to have the value $+\infty$ (though never the value $-\infty$). Dually we denote by C_L the set of all functions h which are the limits of monotone decreasing sequences of functions in C_0. The functional μ can be extended to C_L by means of monotone decreasing limits in the sense that if h is any function in C_L and if (f_n) is any sequence of functions in C_0 which is monotone decreasing with limit h, then we can define $\mu(h)$ uniquely as $\lim_{n\to\infty} \mu(f_n)$.

This time we note that $\mu(h)$ may take the value $-\infty$ but never $+\infty$.

We are now able to offer a general definition of the extended functional μ or, what comes to the same thing, of the extended Stieltjes integral which generates it. For an arbitrary function f we write,

$$\mu^*(f) \equiv \inf\{\mu(g)\}, \quad \text{where } g \in C_U \text{ and } g \geq f.$$

$$\mu_*(f) \equiv \sup\{\mu(h)\}, \quad \text{where } h \in C_L \text{ and } h \leq f.$$

The following properties are easy consequences of the definitions:

(i) $\mu_*(-f) = -\mu^*(f)$,

(ii) $\mu^*(f + g) \leq \mu^*(f) + \mu^*(g)$,

(iii) $\mu_*(f + g) \geq \mu_*(f) + \mu_*(g)$.

We say that f is **widely integrable for** μ if and only if the numbers $\mu^*(f)$ and $\mu_*(f)$ are equal. If in addition their common value is finite then we say that f is **integrable for** μ and write

$$\mu_*(f) = \mu^*(f) \equiv \mu(f) \equiv \int f d\mu. \tag{9.9}$$

9.2.3 Remark. The fact that we are now dealing with functions which may take on infinite values should not cause any difficulty in interpreting pointwise sums such as $f+g$, and so on. We can cope with this situation by adding $+\infty$ and $-\infty$ to the real number system $\mathbb{R}$ as ideal elements and adopting the conventional algebraic rules of the extended real number system $\bar{\mathbb{R}}$, as described in Sec. 1.1.5.

Alternatively, we may tacitly agree to perform arithmetic operations on functions only at those points at which the functions concerned take finite values, leaving combinations like $f + g$ undefined at some points. In the event this often turns out to be unimportant since, in a sense to be explained below, the set of all such points may be described as 'negligible' with respect to the functional μ in question.

9.2.4 If we denote by $L^1(\mu)$ the set of all real, finite-valued functions f which are integrable with respect to a given positive linear functional μ, originally defined on C_0, then the following results may be established:

(i) $L^1(\mu)$ is a linear space, and μ is a non-negative linear functional on $L^1(\mu)$.

(ii) f is integrable for μ if and only if $|f|$ is integrable for μ.

(iii) μ is continuous under monotone limits on $L^1(\mu)$. That is, if the sequence (f_n) of functions in $L^1(\mu)$ converges monotonely to f, then f is at least widely integrable for μ and

$$\mu(f) = \lim_{n \to \infty} \mu(f_n).$$

Apart from this the extension process just described effectively defines μ as a **set function** on a certain family of sets of real numbers. Given any set A of real numbers, we say that the **characteristic function** of A is the function Φ_A defined by

$$\Phi_A(t) = \begin{cases} 1 & \text{for all } t \in A \\ 0 & \text{for all } t \notin A \end{cases}$$

If the characteristic function of A is an integrable function for μ, then A is said to be an integrable set for μ, and we write,

$$\mu(A) \equiv \mu(\Phi_A) = \int \Phi_A d\mu. \tag{9.10}$$

The number $\mu(A)$ is called the μ-**measure** of the set A: if $\mu(A) = 0$ then A is said to be a set of **measure zero for** μ, or a μ-**negligible set**. A property P which holds everywhere except possibly at the points of a set which is of measure zero for μ is said to hold **almost everywhere for** μ. In many applications of integration theory this concept of measure zero is all that is really needed. For completeness, however, we give here a general definition of measurability.

A set of real numbers is said to be **compact** if it is bounded and closed. It can be shown that every compact set K is an integrable set with respect to any (extended) non-negative linear functional μ. Accordingly an arbitrary set A of real numbers is said to be **measurable** for μ if and only if the intersection $A \cap K$ is an integrable set for every compact set K. (Roughly speaking the sets measurable for μ are those which are 'locally integrable' sets for μ).

9.3 THE LEBESGUE AND RIEMANN INTEGRALS

9.3.1 In what follows we confine attention to the particular case of the positive linear functional m defined on C_0 by the ordinary Riemann integral. The resulting extended functional is called the **Lebesgue integral** and the associated set function $m(A)$ is called **Lebesgue measure**. In particular, if A is an interval, $A = [a, b]$ or $A = (a, b)$, then $m(A) = b - a$ so that Lebesgue measure coincides with the ordinary concept of the *length* of the interval. For convenience in what follows we may use any one of the following equivalent forms of notation

$$m(f) \equiv \int f \, dm = \int_{-\infty}^{+\infty} f(t) dt \equiv \int f. \qquad (9.11)$$

The Lebesgue integral, as defined above, is not confined to bounded integrands and is here taken over the whole range from $-\infty$ to $+\infty$. Integration over a finite range $[a, b]$ can be treated within the context of the same general theory by simply noting that

$$\int_a^b f(t) dt = \int_{-\infty}^{+\infty} f(t) \Phi_{[a,b]}(t) dt \equiv \int f \Phi_{[a,b]},$$

where $\Phi_{[a,b]}$ denotes the characteristic function of the interval $[a, b]$.

Now by a **simple function** on $[a, b]$ we shall mean any function s which satisfies the following conditions:

(a) s vanishes identically outside $[a, b]$,

(b) there exists a finite set of points $\{t_k\}$, $1 \leq k \leq n$, such that

$$a = t_0 < t_1 < t_2 < \ldots < t_n = b,$$

and, $s(t) = \alpha_k$ (a finite constant) for $t_{k-1} < t < t_k$.

As is easy to verify, a simple function s_L belongs to the class C_L if and only if at each point of discontinuity, t_k, we have

$$s_L(t_k) \geq \max\{\alpha_k, \alpha_{k+1}\}.$$

Similarly a simple function s_U belongs to the class C_U if and only if we have

$$s_U(t_k) \leq \min\{\alpha_k, \alpha_{k+1}\}.$$

Every simple function s has a well-defined integral over $[a, b]$ which is independent of the values which it assumes at its points of discontinuity.

Now the Riemann integral of a bounded function f over a finite closed interval $[a, b]$ can be defined in terms of the integrals of the simple functions on $[a, b]$. (Without loss of generality we may assume here that f vanishes identically outside $[a, b]$.) First we define the upper and lower Riemann integrals of f over $[a, b]$ as follows:

$$R\left\{\int_a^b f(t)dt\right\}_U = \inf_{s_U(t) \geq f(t)} \int_a^b s_U(t)dt$$

$$R\left\{\int_a^b f(t)dt\right\}_L = \sup_{s_L(t) \leq f(t)} \int_a^b s_L(t)dt$$

It will be noted that in defining the lower integral we use simple functions belonging to C_L whereas for the upper integral we use simple functions belonging to the class C_U. It is easily confirmed that this restriction in no way affects the values of the lower and upper Riemann integrals. However it does allow us to conclude at once that

$$R\left\{\int_a^b f(t)dt\right\}_L \leq m_\star(f) \leq m^\star(f) \leq R\left\{\int_a^b f(t)dt\right\}_U .$$

Since f is Riemann integrable over $[a, b]$ if and only if its upper and lower integrals have the same value, we have the following result:

Let f be a bounded function which is integrable in the Riemann sense over the interval $[a, b]$. Then f is also integrable in the Lebesgue sense over that interval, and the Riemann and Lebesgue integrals have the same value.

Hence without any ambiguity, we can write

$$\int_a^b f(t)dt$$

there being no need to indicate in which sense the integral is defined.

9.3.2 The decomposition of an arbitrary function into positive and negative components (Sec. 8.2.1) continues to hold, with minor modifications, even if f is allowed to take infinite values. We have,

$$f(t) = f^+(t) - f^-(t) \text{ at all points at which } f(t) \text{ is finite,}$$

and

$|f(t)| = f^+(t) - f^-(t)$ everywhere.

From the definition of the Lebesgue integral given above it follows that if f is Lebesgue integrable then so also are the functions f^+ and f^-. This confirms that the Lebesgue theory yields an **absolute** integral in the sense that the integrability of f in the Lebesgue sense always implies the integrability of the function $|f|$. This shows at once that an improper Riemann integral which is only conditionally convergent cannot exist in the Lebesgue sense. Thus, for example, the function $\sin(t)/t$ is not Lebesgue integrable over $(-\infty, +\infty)$ although the improper Riemann integral

$$\int_{-\infty}^{+\infty} \frac{\sin(t)}{t} dt$$

converges to the value π.

Suppose, on the other hand, that the (bounded) function f has an absolutely convergent improper Riemann integral from $-\infty$ to $+\infty$. Without loss of generality we may assume that $f(t) \geq 0$ everywhere. For $n = 1, 2, \ldots$, define

$$f_n(t) = f(t)\Phi_{[-n,n]}(t) = \begin{cases} f(t) & \text{for } -n \leq t \leq n \\ 0 & \text{otherwise} \end{cases} \qquad (9.1)$$

Then,

$$R\int_{-\infty}^{+\infty} f(t)dt = \lim_{n\to\infty} R\int_{-n}^{+n} f(t)dt = \lim_{n\to\infty} \int f_n = \int f.$$

A similar argument can be applied to absolutely convergent improper Riemann integrals of the second kind. Hence we can conclude that

If an improper Riemann integral (of the first, second or third kind) is absolutely convergent then the corresponding Lebesgue integral also exists and has the same value.

9.3.3 So far as the Riemann integral is concerned the immediate justification for developing the alternative Lebesgue theory would seem to be the failure of the former to assign values to the integrals of certain functions. The example quoted in Sec. 9.1.3 concerned a sequence of functions

$(f_m)_{m \in \mathbb{N}}$ where

$$f_m(t) = \lim_{n \to \infty} [\cos(m!\pi t)]^{2n} \quad \text{and} \quad \int_0^1 f_m(t)dt = 0.$$

In view of the discussion in Sec. 9.3.1 there is no need to state in which sense these integrals are to be understood, since the Riemann and Lebesgue integrals of the functions f_m will be identical. For the limit function f of the f_m we have seen that the Riemann integral does not exist. However we have only to note that if $f_m(t) = 1$ for some m then $t = p/q$, where $m \geq q$. Thus, if $n \geq m$ we must have $n \geq q$ and so $f_n(t) = 1$ also. Hence, the sequence $(f_m)_{m \in \mathbb{N}}$ is monotone increasing. By the continuity of the Lebesgue integral under monotone limits the Lebesgue integral from 0 to 1 of the limit function f does exist, and in fact we have

$$\int_0^1 f(t)dt = \lim_{m \to \infty} f_m(t)dt = 0.$$

In point of fact the main practical importance of the Lebesgue integral lies not so much with the resolution of problems such as this as with the much more powerful convergence theorems which can be established for it. As an immediate consequence of the monotone convergence property of the Lebesgue integral we have the following result on termwise integration:

Levi's Theorem. Let $(f_n)_{n \in \mathbb{N}}$ be a sequence of Lebesgue integrable functions such that

$$f_n \geq 0 \quad \text{and} \quad \sum_{k=1}^{n} \int f_k \leq M < +\infty, \quad \text{for all } n.$$

Then the function

$$f(t) \equiv \sum_{k=1}^{+\infty} f_k(t)$$

is Lebesgue integrable over $(-\infty, +\infty)$ and

$$\int_{-\infty}^{+\infty} f(t)dt \equiv \int_{-\infty}^{+\infty} \left\{ \sum_{k=1}^{+\infty} f_k(t) \right\} dt = \sum_{k=1}^{+\infty} \left\{ \int_{-\infty}^{+\infty} f_k(t)dt \right\}.$$

However the most important and powerful of the convergence theorems which can be established for the Lebesgue integral is the so-called **Dominated Convergence theorem** which we quote here (without proof):

Dominated Convergence Theorem. Let $(f_n)_{n\in\mathbb{N}}$ be a sequence of Lebesgue integrable functions which converges almost everywhere to a limit function f. If there exists a Lebesgue integrable function g such that $|f_n| \le g$ for all n then f is Lebesgue integrable and

$$\int_{-\infty}^{+\infty} f(t)dt = \lim_{n\to\infty} \int_{-\infty}^{+\infty} f_n(t)dt.$$

9.3.4 By contrast consider the case of the Riemann-Stieltjes integral in which the integrator is a unit step function $v(t) = u_c(t)$. We know that for any function f continuous on some neighbourhood of the origin (and therefore certainly for any function belonging to C_0) this integral defines a possible model for the delta function described by Dirac. That is, it defines a positive linear functional δ on C_0 given by the mapping

$$f \to \delta(f) = f(0) \equiv \int_{-\infty}^{+\infty} f(t)du_c(t),$$

and we recall that this result is quite independent of the number c. If we now carry out the extension of the functional δ, as described above, the restriction to functions continuous at the origin no longer applies. Consider, for example, the step function $u_0(t)$. This is the characteristic function of the open interval $(0, +\infty)$ and has a jump discontinuity at the origin. But it is easy to see that it is the limit of a monotone increasing sequence of functions f_m belonging to C_0 such that for each m we have $\delta(f_m) = 0$ (see, for example, Fig. 9.1). Hence we must have

$$\int_{-\infty}^{+\infty} u_0(t)du_c(t) = \lim_{m\to\infty} \int_{-\infty}^{+\infty} f_m(t)du_c(t) = 0$$

independently of c.

Similarly we can show that

$$\int_{-\infty}^{+\infty} u_1(t)du_c(t) = 1.$$

The extended δ-functional δ_a (or the δ distribution at a) yields a measure on $\mathbb{R}$ called the **Dirac measure located at** a.

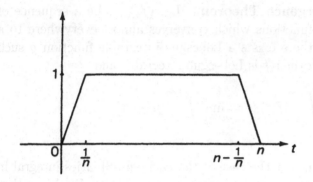

Figure 9.1:

For an arbitrary set $A \subset \mathbb{R}$ we have

$$\delta_a(A) = 1 \ \text{ if } \ a \in A$$

and

$$\delta_a(A) = 0 \ \text{ if } \ a \notin A.$$

Further reading.

Some account of the limitations of the Riemann integral and of the existence of the more flexible and comprehensive theory of the Lebesgue integral ought to form part of any analysis course. The sketch given here is based on the general approach to integration developed by **P.Daniell**. It offers a straightforward and relatively uncluttered way of introducing the Lebesgue integral without having to establish a detailed account of measure theory first. For a comprehensive and eminently readable text on this treatment of Lebesgue integration the classic **Integration** by E.McShane, (Princeton, 1947) is still among the best. More recently the revised and updated edition of **Measure Theory and Integration** by **Gearoid de Barra**, (Horwood Publishing, 2003), gives a most thorough and detailed account of the theory of measure and of Lebesgue-Stieltjes integration.

Chapter 10

Introduction to N.S.A

10.1 A NONSTANDARD NUMBER SYSTEM

10.1.1 Introduction. In the first seven chapters of this book a purely practical approach to the definition and use of the delta function and certain other generalised functions is presented. The outline of the theory of distributions in Chapter 8 shows that a rigorous treatment of generalised functions is possible, but also that it is considerably more demanding. What is more, distributions appear not to be enough, as there seems to be an increasing need for a number of different types of generalised function. Things would be considerably simpler if it were possible to treat all such generalised functions as legitimate functions in the proper sense: that is, as well-defined mappings of numbers into numbers. But to do this, even in the case of the delta function, the simplest and most familiar example, we would seem to need access to an extended version of the real number system, one which would include genuine infinitesimals and infinite numbers. Such extensions of the standard real number system $\mathbb{R}$ do exist, and examples have been studied as long ago as the late 19th century. But the rehabilation of the infinitesimal did not really become established until the appearance, in the 1960s, of a radical reformulation of mathematical analysis due essentially to the work of the mathematical logician **Abraham Robinson**, and published in his seminal text **Non-standard Analysis**, North-Holland, (1966). In this reformulation, which is called **Nonstandard Analysis (NSA)**, generalised functions such as the delta function have virtually the same status as ordinary functions and can be treated accordingly. Its essential features and its general significance can be briefly summarised as follows:

In any standard mathematical discipline we deal with some specific universe $\mathcal{U}$ of well defined mathematical objects (e.g. functions, integrals, operators etc.), based on the fundamental set of the real numbers $\mathbb{R}$. In what is called a 'nonstandard' form of that discipline we deal instead with an enlargement, $^*\mathcal{U}$, of the standard mathematical universe $\mathcal{U}$ which is such that,

(i) $^*\mathcal{U}$ is based on an extension $^*\mathbb{R}$ of $\mathbb{R}$ which contains additional ideal or non-standard objects (infinite numbers and infinitesimals) as well as the standard ones, but which, in a certain sense, preserves all the fundamental properties of $\mathbb{R}$.

(ii) All the standard functions, relations and other higher order objects in $\mathcal{U}$ admit nonstandard extensions to corresponding objects in $^*\mathcal{U}$ which, again in a certain sense, preserve all their fundamental properties.

(iii) There is a **Transfer Principle** in that true statements (theorems) about standard objects in $\mathcal{U}$ can be transferred to form true statements about corresponding nonstandard objects in $^*\mathcal{U}$.

Then there is a pay-off in that useful results about the standard objects of ordinary mathematics can be derived by working in terms of nonstandard objects and exploiting the much richer structure of the 'nonstandard universe', $^*\mathcal{U}$.

In the present chapter a simple form of NSA will be described, based on a particular extension of the usual real number system, $\mathbb{R}$, which is defined in terms of equivalence classes of real sequences. This is an enlarged copy $^*\mathbb{R}$ of the real number system $\mathbb{R}$ which contains both infinitesimal and infinite elements as well as (copies of) the real numbers themselves. The term 'copy' is used advisedly here since, in a sense which will be explained more fully later, $^*\mathbb{R}$ does inherit all the properties of $\mathbb{R}$. In particular we can handle and manipulate the infinitesimal and infinite elements of $^*\mathbb{R}$ using the same algebraic rules as those which apply to ordinary real numbers. It will be used subsequently to give a genuinely elementary but entirely rigorous interpretation of the generalised functions which have been treated on a purely formal basis in the earlier chapters of this book.

$^*\mathbb{R}$ is called a system of **hyperreal numbers**. Each ordinary, standard, function f defined on $\mathbb{R}$ has a canonical 'nonstandard' extension to a function $^*f$ defined on (and taking values in) $^*\mathbb{R}$. It will be seen that such an extension $^*f$, does inherit, in a certain sense, all the properties of f. This allows in particular a development of elementary calculus along Leibnitzian lines with a free use of arguments based on infinitesimals rather than on the more orthodox (but conceptually less attractive) concept of limit. There is a resulting gain in simplicity but no loss of rigour. In a justly celebrated text by **H.Jerome Keisler**, called **Elementary Calculus: An infinitesimal approach** (1986), such an exposition of elementary calculus is developed from first principles and has been successfully used to teach calculus to beginners.

Although it is comprehensive Robinson's original book on NSA is far from easy and demands some expertise in mathematical logic and model theory. Simpler presentations of NSA have since been developed, in particular a 'constructive' approach in which a model of a nonstandard universe can be exhibited as a so-called ultrapower. We begin by sketching an elementary form of such a model of NSA which is adequate for a nonstandard theory of the delta function and of other generalised functions.

10.1.2 Hyperreal numbers. The basic idea is to define an enlargement of the ordinary real number system $\mathbb{R}$ in terms of equivalence classes of infinite real sequences $(x_n)_{n \in \mathbb{N}}$. Consider the set $\mathbb{R}^{\mathbb{N}}$ of all such sequences. We could equip this set with an algebraic structure in a natural way by introducing componentwise definitions of addition and multiplication:

$$(x_n)_{n \in \mathbb{N}} + (y_n)_{n \in \mathbb{N}} = (x_n + y_n)_{n \in \mathbb{N}}$$

$$(x_n)_{n \in \mathbb{N}} \times (y_n)_{n \in \mathbb{N}} = (x_n \times y_n)_{n \in \mathbb{N}}.$$

This would ensure that such addition and multiplication would obey the usual laws of arithmetic (associative, commutative and distributive), and that there would exist additive and multiplicative unit elements defined respectively by the null sequence and the unit sequence:

$$(0, 0, 0, ...) \quad \text{and} \quad (1, 1, 1, ...).$$

If we agree to identify each real number x with the constant sequence in which $x_n = x$ for all n then it is easy to see that we can embed $\mathbb{R}$ (together with its usual arithmetic structure) in $\mathbb{R}^{\mathbb{N}}$. This would certainly give us a number system of sorts which is an enlargement of $\mathbb{R}$, but it would have many undesirable features. For one thing it would be much too large. Even a sequence like $(1, 0, 0, 0, ...)$, which has only one non-zero entry, would have to be regarded as a number distinct from zero.

Some improvement would result if instead of taking individual sequences $(x_n)_{n \in \mathbb{N}}$ as the elements of the proposed extended number system we used suitably defined equivalence classes of such sequences. An obvious first choice might be to require that two sequences $(x_n)_{n \in \mathbb{N}}$ and $(y_n)_{n \in \mathbb{N}}$ should be equivalent if $x_n = y_n$ for all but finitely many values of n. Then, for example, the sequence $(1, 0, 0, 0, ...)$ would simply be another representative member of the equivalence class standing for the null, or zero, element of the system as also would such sequences as $(2, 0, 0, 0, ...)$ or $(2, 3, 0, 0, 0, ...)$. In a remarkable anticipation of some aspects of Nonstandard Analysis proper, an extension of the real number system using equivalence classes defined in this way was published in 1958 by **C.Schmieden** and **D.Laugwitz** (cf. **Eine Erweiterung des Infinitesimalkalkuls**, Math. Z., vol. 69, pp. 1-39). The resulting number system was an extension of $\mathbb{R}$ which certainly contained elements which could be identified as genuine infinitesimals and infinite numbers, and which was used by the authors to present an intuitively appealing interpretation of delta functions and to develop an appropriate delta function calculus.

Nevertheless this number system has its drawbacks: it is not a field, and it contains divisors of zero. Consider, for example, two such sequences as $(0, 1, 0, 1, 0, 1, ...)$ and $(1, 0, 1, 0, 1, 0, ...)$. They are clearly not equivalent in the required sense and therefore represent two distinct equivalence classes and so two distinct elements of the system. Moreover they have a componentwise product given by the null sequence,

$$(0, 1, 0, 1, 0, 1, ...) \times (1, 0, 1, 0, 1, 0, ...) = (0, 0, 0, 0, 0, ...)$$

although neither of the factor sequences is null.

To eliminate such divisors of zero and obtain a genuine field we need a criterion for equivalence which is consistent with, but more comprehensive than, the condition that x_n should equal y_n for all but finitely many values of n. Accordingly we proceed as follows:

We shall say that $(x_n)_{n \in \mathbb{N}}$ and $(y_n)_{n \in \mathbb{N}}$ are equivalent whenever $x_n = y_n$ for **nearly all** n, where the term 'nearly all' is to have an extended meaning which satisfies the following conditions:

NA(1) If A is any subset of $\mathbb{N}$ then either A contains 'nearly all' members of $\mathbb{N}$ or else its complement $\mathbb{N} - A$ does.

NA(2) If A is any finite subset of $\mathbb{N}$ then $\mathbb{N} - A$ contains 'nearly all' members of $\mathbb{N}$.

NA(3) If A and B are subsets of $\mathbb{N}$ which each contain 'nearly all' members of $\mathbb{N}$ then so also does $A \cap B$.

NA(4) If A contains 'nearly all' members of $\mathbb{N}$ and if $A \subset B \subset \mathbb{N}$, then B also contains 'nearly all' members of $\mathbb{N}$.

We denote by $^*\mathbb{R} \equiv \mathbb{R}^{\mathbb{N}}/\sim$ the corresponding set of equivalence classes. The equivalence class containing a sequence (x_n) will be written as $[(x_n)_{n \in \mathbb{N}}]$, or sometimes simply as $[x_n]$, and will be called a **hyperreal number**. Thus, if $x \equiv [x_n]$ is the hyperreal number defined by the equivalence class containing the sequence $(x_n)_{n \in \mathbb{N}}$ then for any other sequence $(y_n)_{n \in \mathbb{N}}$ we will have

$(y_n)_{n \in \mathbb{N}} \in [x_n]$ if and only if $y_n = x_n$ for nearly all $n \in \mathbb{N}$.

$^*\mathbb{R}$ is equipped with algebraic operations and an ordering relation in the natural way:

Addition: $[z_n] = [x_n] + [y_n]$ if $x_n + y_n = z_n$ for nearly all n.

Multiplication: $[z_n] = [x_n] \times [y_n]$ if $x_n \times y_n = z_n$ for nearly all n.

Order: $[x_n] < [y_n]$ if $x_n < y_n$ for nearly all n.

10.1.3 Remark. It can be shown that assigning such an extended meaning to the term 'nearly all' is equivalent to fixing a finitely additive Boolean measure m on the set $\mathbb{N}$ of natural numbers. That is to say it effectively assumes the existence of a function m mapping $\mathbb{N}$ into the set $\{0, 1\}$ which is such that

M(1) For each $A \subset \mathbb{N}$ we have either $m(A) = 0$ or else $m(A) = 1$.

M(2) If $A \cap B = \emptyset$ then $m(A \cup B) = m(A) + m(B)$; (finite additivity).

M(3) $m(A) = 0$ for every finite subset A of $\mathbb{N}$.

Then $(x_n)_{n \in \mathbb{N}}$ is equivalent to $(y_n)_{n \in \mathbb{N}}$ if and only if the set $\{n \in \mathbb{N} : x_n = y_n\}$ has measure 1. We cannot give an explicit characterisation of such a measure m since this would involve stating infinitely many special conditions to cover all possible sequences. Moreover there could in any case be no unique such characterisation. Surprisingly, perhaps, this turns out not to matter. It is enough to know merely that such a measure exists in order to resolve the problem of the divisors of zero. We know for example that either the equivalence class $[(0, 1, 0, 1, 0, \ldots)] = 0$ and $[(1, 0, 1, 0, 1, \ldots)] = 1$ or else that $[(0, 1, 0, 1, 0, \ldots)] = 1$ and $[(1, 0, 1, 0, 1, \ldots)] = 0$. Which alternative actually obtains depends on whether we were to define the measure m such that $m(\{1, 3, 5, 7, \ldots\}) = 1$ and $m(\{2, 4, 6, 8, \ldots\}) = 0$ or instead such that $m(\{1, 3, 5, 7, \ldots\}) = 0$ and $m(\{2, 4, 6, 8, \ldots\}) = 1$. In the event it is enough to know that just one of these two possibilities must hold.

10.1.4 Structure of $^*\mathbb{R}$. $^*\mathcal{R}$ has the same algebraic properties as $\mathbb{R}$ itself. That is, the definitions of addition, multiplication and order given above ensure that the axioms of an ordered field are satisfied for $^*\mathbb{R}$. Moreover $\mathbb{R}$ is embedded in $^*\mathbb{R}$ as a sub-ordered field by identifying each $x \in \mathbb{R}$ with the equivalence class $[x]$ in $^*\mathbb{R}$ which contains the constant sequence $x_n = x$ for all n. The real numbers, as members of $^*\mathbb{R}$, are called the **standard elements** of $^*\mathbb{R}$. Among the other, nonstandard, elements of $^*\mathbb{R}$ there will be those which play the roles of infinite numbers and of infinitesimal numbers respectively. In fact we have the following classification of the general (standard and nonstandard) elements of $^*\mathbb{R}$:

Infinite numbers. An element $x = [x_n]$ of $^*\mathbb{R}$ is an **infinite hyperreal number** if for every positive standard (i.e. real) number r we have $|x| > r$; that is to say, if for every $r \in \mathbb{R}_+$ the inequality $|x_n| > r$ holds for nearly all values of n in $\mathbb{N}$. The set of all infinite hyperreal numbers will be denoted by $^*\mathbb{R}_\infty$.

As a typical example of an infinite hyperreal, take the element Ω defined by the equivalence class

$$\Omega \equiv [n] = [(1, 2, 3, 4, \ldots)]. \tag{10.1}$$

Infinitesimal numbers. An element $x = [x_n]$ of $^*\mathbb{R}$ is an **infinitesimal hyperreal number** if for every positive standard number r we have $|x| \leq r$: that is to say, if for every $r \in \mathbb{R}_+$ the inequality $|x_n| \leq r$ holds for nearly all values of n in $\mathbb{N}$. The set of all infinitesimal numbers will be denoted by $^*\mathbb{R}_0$.

As a typical example of an infinitesimal hyperreal we may take the element ε, defined as the equivalence class

$$\varepsilon \equiv [1/n] = [(1, 1/2, 1/3, 1/4, \ldots)]. \tag{10.2}$$

If r is any given positive real number then for all sufficiently large values of n we will have

$$0 < 1/n < r$$

so that ε does behave as one would expect of a positive infinitesimal element. It is, of course, the multiplicative inverse of the infinite hyperreal number Ω cited above, for we have

$$\Omega \times \varepsilon = [(1, 2, 3, 4, \ldots)] \times [(1, 1/2, 1/3, 1/4, \ldots)]$$

$$= [(1, 1, 1, 1, \ldots)] = 1.$$

As one might expect, the sum of two infinitesimals is infinitesimal and the product of any finite number with any infinitesimal is an infinitesimal.

Finite numbers. An element $x = [x_n]$ of $^*\mathbb{R}$ is a **finite hyperreal number** if there exists a positive standard number r such that $|x| \leq r$: that is to say, if for some $r \in \mathbb{R}_+$ the inequality $|x_n| \leq r$ holds for nearly all values of n in $\mathbb{N}$. The set of all finite hyperreal numbers will be denoted by $^*\mathbb{R}_{bd}$. All the standard elements of $^*\mathbb{R}$ (that is, all ordinary real numbers) are members of $^*\mathbb{R}_{bd}$, and so also are all infinitesimals.

$^*\mathbb{R}_{bd}$ is not a field since not every element x in $^*\mathbb{R}_{bd}$ has an inverse x^{-1} which belongs to $^*\mathbb{R}_{bd}$. (If ξ is any non-zero infinitesimal then $\xi \in {}^*\mathbb{R}_{bd}$ but $\xi^{-1} \in {}^*\mathbb{R}_\infty$.) On the other hand it contains no divisors of zero. Technically the set $^*\mathbb{R}_{bd}$ is a **commutative ring with identity** and $^*\mathbb{R}_0$ is a **maximal proper ideal** of that ring.)

10.1.5 Remarks. (i) The hyperreal number system $^*\mathbb{R}$ has been said (loosely) to inherit 'all the properties of the ordinary real number system $\mathbb{R}$'. This statement is true but only in so far as the first order properties of $\mathbb{R}$ are concerned: that is, only for those properties which concern numbers themselves and their arithmetic. It is not generally true with respect to properties of sets of numbers. For example the fact that $\mathbb{R}$ is a totally ordered field is a first order property of $\mathbb{R}$ and, as already remarked, it is also the case that $^*\mathbb{R}$ is a totally ordered field. This is a property which transfers directly from $\mathbb{R}$ to $^*\mathbb{R}$.

On the other hand consider the fact that $\mathbb{R}$ is an Archimedean field. This is not a first order property of $\mathbb{R}$, and it does not transfer to $^*\mathbb{R}$ which contains infinitesimal elements and so is quite plainly not Archimedean.

Similarly, while it is true that the real numbers form a complete orderd field, the hyperreal number system is not complete. For $\mathbb{R}$ satisfies the fundamental least upper bound property which can be stated as follows:

Every non-empty subset A of $\mathbb{R}$ which is bounded above in $\mathbb{R}$ has a least upper bound in $\mathbb{R}$.

This is another property which is not first order and which does not transfer to $^*\mathbb{R}$, at least in any direct sense. Thus, $\mathbb{R}$ is itself a non-empty subset of $^*\mathbb{R}$ which is bounded above in $^*\mathbb{R}$ (by any positive infinite element of $^*\mathbb{R}$). But it clearly has no least upper bound in $^*\mathbb{R}$ since if Λ is any positive infinite hyperreal then so also is $\Lambda - 1$. To obtain valid analogues of this, and other such properties, in $^*\mathbb{R}$ we shall need to introduce in due course a special classification of the subsets of $^*\mathbb{R}$ into two categories; these are the so-called **internal** and **external** sets. (See section 10.4 below).

(ii) A word of caution on the notation may be in order here. The symbol ε is frequently used in standard mathematics to denote a small positive real number. Throughout what follows we will always reserve for it the meaning given by (10.2); that is, ε will always represent in future a specific positive infinitesimal. It would be nice to adhere to a consistent usage which always distinguished between standard real numbers and the various nonstandard hyperreals which we will be using from time to time. This has not proved possible, but the context should make the distinction clear.

10.1.6 The hyperreal number line. A quasi-geometric picture of a hyperreal number line is sometimes offered in the form of an extended version of the usual illustration of the real number line. (Fig. 10.1) The finite part of the hyperreal line appears in the centre of such a diagram looking, it must be confessed, very much like the familiar picture of the real number line itself. This is not too surprising. The real number system may be regarded as an abstract model of the intuitive ideas which we have about the structure of the ideal straight line of our geometric imagination. If we take any point of this conceptual line then we expect to be able to assign to it a specific real number. There seems to be no room to fit in any additional nonstandard numbers, and it is by no means clear in what way we might be able to locate a point which has been labelled by some finite, but nonstandard hyperreal. In this sense infinitesimals (and finite nonstandard hyperreals in general) might be considered counter-intuitive.

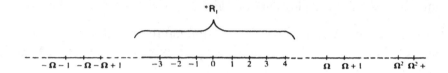

Figure 10.1:

An ingenious solution to the problem of visualising the finite part of the hyperreal line was proposed originally by K.D.Stroyan and subsequently exploited by Jerome Keisler (op. cit.) and others. (Fig. 10.2) In this portions of the hyperreal line are supposedly blown up, or expanded, by means of an 'infinitesimal microscope'.

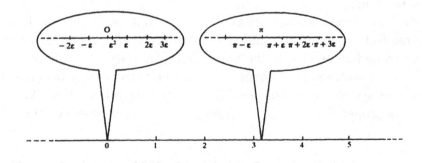

Figure 10.2:

This useful instrument is assumed to have the power of magnifying to an infinite order. As a result we are able to 'see' points infinitely close to specific real numbers, which are not normally visible to the unaided eye of our imagination.

10.1.7 Monads and standard parts. Two hyperreals x, y are said to be **infinitely close** if the difference $(x - y)$ is infinitesimal, and we then write $x \approx y$. (The binary relation $\approx$ is an equivalence relation on the ring $^*\mathbb{R}_{bd}$ of all finite elements of $^*\mathbb{R}$.)

For each finite x in $^*\mathbb{R}$ there exists a unique real number r with $x \approx r$; this real number r is called the **standard part** of x and is written as $\text{st}(x)$, or sometimes as $^o x$. The set of all hyperreal numbers which are infinitely close to a given real number r is called the **monad** of r:

$$\text{mon}(r) := \{x \in {}^*\mathbb{R} : \ \text{st}(x) = r\} \equiv \{x \in {}^*\mathbb{R} : \ x \approx r\}. \qquad (10.3)$$

The quotient set $^*\mathbb{R}_{bd}/ \approx$ consisting of the equivalence classes of the finite hyperreals with respect to the equivalence relation $\approx$ turns out to be just the ordinary real number system $\mathbb{R}$ itself. In passing from $\mathbb{R}$ to $^*\mathbb{R}_{bd}$ we effectively replace each real number r by an entire set of finite hyperreals, each one of which is infinitely close to r. In a more picturesque sense the infinitesimal microscope of the preceding section can be thought of as expanding each real (i.e. standard) point of the line into its monad.

In what follows we shall be concerned to develop mathematical analysis in a universe based on a hyperreal number system $^*\mathbb{R}$ rather than in the familiar universe based on the ordinary real number system $\mathbb{R}$. The higher order mathematical objects encountered in the standard universe (sets of numbers, functions, integrals, etc.) will be extended to corresponding objects in the nonstandard universe, as explained in the following sections. In passing it should be noted that a hypercomplex number system $^*\mathbb{C}$ can be defined in an obvious way as $^*\mathbb{C} = {}^*\mathbb{R} + i\,^*\mathbb{R}$. It is a proper extension of the standard complex number system $\mathbb{C}$.

10.2   NONSTANDARD EXTENSIONS

10.2.1 Extension of a set. If A is any standard subset of $\mathbb{R}$ then we define its **nonstandard extension** to be the set $^*A$ of all hyperreals

$x = [x_n]$ such that $x_n \in A$ for nearly all n,

$$^*A = \{x = [x_n] : \ x_n \in A \ \text{for nearly all} \ n\}. \tag{10.4}$$

If A contains only finitely many members then we will have $A = {}^*A$, but in general ${}^*A$ will be a proper extension of A.

Examples.

(i) The set $\mathbb{N}$ of all natural numbers is a subset of $\mathbb{R}$ which has a nonstandard extension ${}^*\mathbb{N}$ which we call the set of all **hypernatural numbers**. Each member of this set ${}^*\mathbb{N}$ is an equivalence class $[(m_n)_{n \in \mathbb{N}}]$ where the m_n are natural numbers (positive integers). ${}^*\mathbb{N}$ will contain infinite elements, such as $\Omega = [n]$, as well as (copies of) the natural numbers themselves:

$$^*\mathbb{N} = \{1, 2, 3, \dots, \Omega - 1, \Omega, \Omega + 1, \dots, \Omega^2, \Omega^2 + 1, \dots\}$$

It will sometimes be convenient to denote by ${}^*\mathbb{N}_\infty$ the set ${}^*\mathbb{N} - \mathbb{N}$ of all infinite hypernatural numbers.

(ii) If $[a, b]$ is any real (finite) interval in $\mathbb{R}$ then its nonstandard extension ${}^*[a, b]$ is a hyperreal interval in ${}^*\mathbb{R}$ containing not only all the real numbers belonging to $[a, b]$, but *all* hyperreal numbers $x = [x_n]$ such that $a \leq x_n \leq b$ for (nearly) all n. Thus, the hyperreal interval ${}^*[0, 1]$ contains in particular, all positive infinitesimal elements ξ as well as the standard number 0, and all hyperreal numbers $(1 - \xi)$ as well as the standard number 1.

Finally note that the set ${}^*\mathbb{R}$ of all hyperreal numbers is, of course, the nonstandard extension of the set $\mathbb{R}$ of all real numbers.

10.2.2 Extension of a function. We can now define the canonical extension of an arbitrary standard function f.

If f is a standard function defined on a set $A \subset \mathbb{R}$ then its nonstandard extension is the function ${}^*f$ defined on the set of hyperreals ${}^*A$ which is given by the rule

$$^*f(x) = [f(x_n)], \ \text{for each} \ x = [x_n] \in {}^*A. \tag{10.5}$$

The restriction of ${}^*f$ to A is sometimes called the **shadow** of ${}^*f$, and it coincides with the original function f on A.

Examples.

(i) The standard function $\sin(r)$ is well defined for all real numbers r: its nonstandard extension is a function $^*\sin(x)$ which is well defined on the whole of $^*\mathbb{R}$. At any standard point, say $x = r$, we have $^*\sin(r) = \sin(r)$. Elsewhere, for example, we have

$$^*\sin(\varepsilon) \equiv [\sin(1/n)] \approx 0, \quad \text{and} \quad ^*\sin(\Omega\pi) \equiv [\sin(n\pi)] = 0.$$

(ii) The standard function $f(r) = 1/r$ is well defined on the set $A = \mathcal{R} - \{\prime\}$. Its nonstandard extension $^*f(x) = x^{-1}$ is defined on $^*\mathbb{R} - \{0\}$: it takes infinite hyperreal values when x is a non-zero infinitesimal, infinitesimal values when x is infinite, and coincides with its shadow $f(r) = 1/r$ for all real non-zero values $x = r$.

The nonstandard extension $^*f$ of a standard function f exhibits all the algebraic properties of f. Thus, for example, we have

$$^*\sin^2(x) + {}^*\cos^2(x) = 1,$$

$$^*\sin(x + y) = {}^*\sin(x)\,{}^*\cos(y) + {}^*\cos(x)\,{}^*\sin(y),$$

and so on. Except when it is necessary to emphasise the fact that we are dealing with a nonstandard extension it is usual to omit the asterisk and use the same familiar notation for both the standard function and its nonstandard extension.

10.3   ELEMENTARY ANALYSIS

10.3.1 The value of NSA at an elementary level is usually illustrated by applying it to elementary calculus and to basic real analysis. We do so here, very briefly, to indicate some of the essential ideas. Among these the nonstandard treatment of the concept of limit is perhaps the most immediately significant. We consider first a nonstandard approach to the idea of the limit of an infinite sequence of real numbers.

10.3.2 Sequences and series. A real sequence $(a_n)_{n \in \mathbb{N}}$ is, properly speaking, a function defined on the set $\mathbb{N}$ of the positive integers and taking real values. That is, it is a mapping $a : \mathbb{N} \to \mathbb{R}$ with $a(n) \equiv a_n$. As such it has a nonstandard extension, $^*a$, which is defined on the set

*N and which takes hyperreal values. This mapping agrees with the original sequence on ℕ, so that we have, for any (standard) positive integer n,

$$a_n \equiv a(n) = {}^*a(n) \equiv {}^*a_n.$$

Now let Λ be any infinite hypernatural number. Then Λ is an equivalence class of the form $\Lambda = [(m_n)_{n \in \mathbb{N}}]$, where $(m_n)_{n \in \mathbb{N}}$ is a sequence of positive integers which is unbounded above. The Λth term of the extended sequence is the hyperreal number given by

$$ {}^*a(\Lambda) \equiv a_\Lambda = [(a_{m_n})_{n \in \mathbb{N}}].$$

For example, let $a_n = 1/2^{n-1}$ for $n \in \mathbb{N}$. Then, recalling that Ω is the infinite hypernatural number defined by the equivalence class $[n]$, we can take $\Lambda = \Omega,\ 2\Omega,\ 2\Omega - 1,\ \Omega^2$, and so on to get

$$a_\Omega = [(a_n)_{n \in \mathbb{N}}] = [(1, 1/2, 1/4, 1/8, \ldots)],$$

$$a_{2\Omega} = [(a_{2n})_{n \in \mathbb{N}}] = [(1/2, 1/8, 1/32, 1/128, \ldots)],$$

$$a_{2\Omega-1} = [(a_{2n-1})_{n \in \mathbb{N}}] = [(1, 1/4, 1/16, 1/64, \ldots)],$$

$$a_{\Omega^2} = [(a_{n^2})_{n \in \mathbb{N}}] = [(1, 1/8, 1/256, \ldots)].$$

Convergence. Suppose that $(a_n)_{n \in \mathbb{N}}$ is a real sequence which converges to the real number α as its limit. Then, given any positive real number r there must exist some positive integer $n_0 = n_0(r)$ such that

$$|\alpha - a_n| < r$$

for all $n \geq n_0$, and therefore for nearly all $n \in \mathbb{N}$. Now let $\Lambda = [(m_n)_{n \in \mathbb{N}}]$ be any infinite hypernatural number. Since Λ is infinite we know that $m_n \geq n_0$ for nearly all $n \in \mathbb{N}$, and it follows that

$$|\alpha - a_{m_n}| < r \text{ for nearly all } n \in \mathbb{N}.$$

Since r is arbitrary this means that we must have $\alpha - a_\Lambda \approx 0$.

Suppose on the other hand that $(a_n)_{n \in \mathbb{N}}$ does not converge to α. Then there must exist some real $r > 0$ such that for each $n \in \mathbb{N}$ we can find an integer $m_n \geq n$ for which we have $|\alpha - a_{m_n}| \geq r$. The sequence $(m_n)_{n \in \mathbb{N}}$ defines a hypernatural number Λ, necessarily infinite, for which $|\alpha - a_\Lambda| \geq r$, and therefore such that a_Λ cannot be infinitely close to α.

Thus, we have the following nonstandard criterion for convergence of a real sequence:

Proposition 10.1. $\lim_{n \to \infty} a_n = \alpha \in \mathbb{R}$ if and only if $a_\Lambda \approx \alpha$ for every $\Lambda \in {}^*\mathbb{N}_\infty$.

Infinite series. In standard analysis an infinite series is not an infinite sum but the limit of a sequence of finite sums. We do not try to interpret an expression like $\sum_{k=1}^{+\infty} a_k$ as an instruction to carry out an infinite number of additions. By contrast the nonstandard approach does allow us to talk of literally infinite sums:

For each $n \in \mathbb{N}$ we can define a finite sum $A_n = \sum_{k=1}^{n} a_k$, and the sequence $(A_n)_{n \in \mathbb{N}}$ is itself a mapping A carrying $\mathbb{N}$ into $\mathbb{R}$ in which $A(n) \equiv A_n$. Hence it extends to a mapping ${}^*A$ of ${}^*\mathbb{N}$ into ${}^*\mathbb{R}$, and for any infinite hypernatural number $\Lambda = [(m_n)_{n \in \mathbb{N}}]$ it will make sense to talk of the "sum to Λ terms" since this is well defined as

$$ {}^*A(\Lambda) \equiv A_\Lambda = \sum_{k=1}^{\Lambda} a_k = \left[\left(\sum_{k=1}^{m_n} a_k \right)_{n \in \mathbb{N}} \right]. \qquad (10.6) $$

In standard terminology we say that the infinite series $\sum_{k=1}^{+\infty} a_n$ converges to the real number s as its sum if and only if

$$ \lim_{n \to \infty} A_n \equiv \lim_{n \to} \sum_{k=1}^{n} a_k = s. $$

According to the nonstandard criterion for convergence this will be the case if and only if for every infinite hypernatural number Λ we have

$$ \sum_{k=1}^{\Lambda} a_k \approx s. $$

For example, let $A_n = \sum_{k=1}^{n} 1/k(k+1)$ so that

$$ A_\Lambda = \sum_{k=1}^{\Lambda} \frac{1}{k(k+1)} = \sum_{k=1}^{\Lambda} \left\{ \frac{1}{k} - \frac{1}{k+1} \right\} $$

$$= \sum_{k=1}^{\Lambda} \frac{1}{k} - \sum_{k=1}^{\Lambda} \frac{1}{k+1} = 1 - \frac{1}{\Lambda+1},$$

which is infinitely close to 1. This result, in nonstandard terms, corresponds to the standard statement that the infinite series $\sum_{k=1}^{+\infty} 1/k(k+1)$ converges to the sum 1.

On the other hand, consider infinite sums of the form $\sum_{k=1}^{\Lambda}(-1)^{k+1}$ where $\Lambda \in {}^*\mathbb{N}_\infty$. For $\Lambda = 2\Omega \equiv [2n]$ and $\Lambda = 2\Omega - 1 \equiv [2n-1]$ the values of the corresponding infinite sums are

$$\sum_{k=1}^{2\Omega}(-1)^{k+1} = \left[\left(\sum_{k=1}^{2n}(-1)^{k+1}\right)_{n\in\mathbb{N}}\right] = [0] = 0$$

and

$$\sum_{k=1}^{2\Omega-1}(-1)^{k+1} = \left[\left(\sum_{k=1}^{2n-1}(-1)^{k+1}\right)_{n\in\mathbb{N}}\right] = [1] = 1.$$

The nonstandard criterion cannot therefore be fulfilled, and so the infinite series $\sum_{k=1}^{+\infty}(-1)^{k+1}$ cannot be convergent.

10.3.3 Elementary topology of $\mathbb{R}$. In elementary calculus it is usually enough to distinguish between open and closed intervals without enquiring further into the definition and properties of open and closed sets of real numbers in general. Later on it becomes necessary to study more closely the so-called **topology** of the real line and this is much simpler to deal with in nonstandard terms. In the usual standard treatment the basic idea is that of **neighbourhood**:

A set U of real numbers is said to be a neighbourhood of a point $a \in \mathbb{R}$ if and only if U contains all real numbers x which are 'sufficiently close' to a. More precisely U is a neighbourhood of a if and only if there exists some $r > 0$ such that

$$U \supset \{x \in \mathbb{R} : |x - a| < r\}.$$

It is easy to restate this criterion in nonstandard terms:

If $x = [x_n]$ is any hyperreal which is infinitely close to a then for any real $r > 0$ we must have $|a - x_n| < r$ for nearly all n. Hence if U is a neighbourhood of a then $x_n \in U$ for nearly all n, and so x must be a member of $^*U$. That is to say, if a set U is a neighbourhood of a point $a \in \mathbb{R}$ then we must have $^*U \supset \text{mon}(a)$.

Suppose, on the other hand, that U is any set of reals such that $^*U \supset \text{mon}(a)$. If U were not a neighbourhood of a, then for each $n \in \mathbb{N}$ we could find a point $x_n \in \mathbb{R}$ such that

$$|a - x_n| < 1/n \text{ and } x_n \in \mathbb{R} - U.$$

But then $x = [x_n]$ would be a hyperreal which belongs to $\text{mon}(a)$ but not to $^*U$, contrary to hypothesis. Hence we have the result:

Proposition 10.2. A set $U \subset \mathbb{R}$ is a neighbourhood of a point $a \in \mathbb{R}$ if and only if $\text{mon}(a) \subset {}^*U$.

This nonstandard characterisation of neighbourhood allows us to frame simple, nonstandard definitions of open, closed and compact sets in $\mathbb{R}$:

A subset G of $\mathbb{R}$ is said to be **open** if and only if G is a neighbourhood of each of its points. In nonstandard terms this will be the case if and only if for every real $x \in G$ we have $\text{mon}(x) \subset {}^*G$.

Similarly the standard condition for a $F \subset \mathbb{R}$ to be **closed** is that its complement $\mathbb{R} - F$ should be open. The corresponding nonstandard criterion simply states that F is closed if and only if for every finite hyperreal $x \in {}^*F$ we have $st(x) \in F$.

A subset K of $\mathbb{R}$ is said to be **compact** if and only if it is bounded and closed. It follows immediately that $K \subset \mathbb{R}$ is compact if and only if every hyperreal $x \in {}^*K$ is a finite point such that $st(x) \in K$.

Continuity. Let f be a standard function which is well defined at least on a neighbourhood of some point a in $\mathbb{R}$. Then $^*f(x)$ is certainly well defined for all x in $^*\mathbb{R}$ such that $x \approx a$. The standard condition for the continuity of f at a requires that for all points x sufficiently near a we should have $f(x)$ near $f(a)$. To be more precise we say that f is continuous at a if and only if, given any positive real number r we can always find a corresponding positive real number s (usually dependent on both r and a) such that

$$|f(x) - f(a)| < r \text{ whenever } |x - a| < s.$$

Using the nonstandard characterisation of neighbourhood derived above it is easily confirmed that this is equivalent to the following, more direct, nonstandard criterion:

Proposition 10.3. f is continuous in the standard sense at a point $a \in \mathbb{R}$ if and only if $^*f(x) \approx f(a)$ for all hyperreal $x \approx a$.

That is, f is continuous at $a \in \mathbb{R}$ if and only if

$$\forall x \in {}^*\mathbb{R} : x \in \mathrm{mon}(a) \Rightarrow {}^*f(x) \in \mathrm{mon}(f(a)). \tag{10.7}$$

Properties of standard continuous functions are simple and direct consequences of this nonstandard criterion. For example, if f is continuous at $a \in \mathbb{R}$ and if g is a function which is continuous at the point $f(a) \in \mathbb{R}$ then for all $x \in {}^*\mathbb{R}$ such that $x \approx a$ we have immediately

$$^*g(^*f(x)) \approx g(f(a))$$

so that the composition $g \circ f$ is continuous at $a \in \mathbb{R}$.

Similarly it is easy to show that the following nonstandard criterion for uniform continuity is equivalent to the standard one:

The standard function $f : \mathbb{R} \to \mathbb{R}$ is **uniformly continuous** on $A \subset \mathbb{R}$ if and only if for every pair of hyperreals $x, y \in {}^*A$ such that $x \approx y$ we have $^*f(x) \approx {}^*f(y)$.

Arguably this may seem easier to understand than the standard definition and it is certainly easier to apply. For example, if $f(x) = 1/x$ then f is defined on the set $A = (0, +\infty)$. If $x \in A$, let $y = x + \xi$, where ξ is an infinitesimal. Then $y \in {}^*A$ and we have

$$^*f(x) - {}^*f(y) = 1/x - 1/(x + \xi) = \xi/x(x + \xi)$$

which is surely infinitesimal. Hence f is continuous at x and therefore (since x is arbitrary) on the whole of A. However if instead we take for x and y two positive infinitesimal points, say $\epsilon = [1/n]$ and $\eta = [1/(n + 1)]$ then we would get

$$^*f(\epsilon) - {}^*f(\eta) = -1$$

which is certainly not infinitesimal. This shows, in a very simple and straightforward manner, that f is not uniformly continuous on $(0, +\infty)$.

A nice example of the simplifying power of nonstandard methods can be seen in a nonstandard proof of the theorem that continuity on a compact set (and in particular on any bounded, closed real interval) implies uniform continuity there:

Let f be defined and continuous on a compact set $K \subset \mathbb{R}$. If x and y are any points in $^*K$ such that $x \approx y$ then, since K is compact, there must exist some point $a \in K$ such that

$$st(x) = a = st(y).$$

But by the continuity of f at a we must then have

$$^*f(x) \approx f(a) \quad \text{and} \quad ^*f(y) \approx f(a).$$

Hence $^*f(x) \approx {}^*f(y)$ and it follows at once that f is uniformly continuous on K.

Differentiation. If f is differentiable at $a \in \mathbb{R}$ then its derivative $f'(a)$ is the (unique) real number given by

$$f'(a) = st \left\{ \frac{^*f(x) - f(a)}{x - a} \right\}, \tag{10.8}$$

where x is any hyperreal such that $x \approx a$.

This definition allows a particularly simple derivation of the chain rule, $(f \circ g)'(a) = f'(g(a))g'(a)$. Thus, let $x \approx a$ but $x \neq a$; all we have to prove is that

$$st \left\{ \frac{f(g(x)) - f(g(a))}{x - a} \right\} = f'(g(a))g'(a).$$

If $g(x) = g(a)$ then $g'(a) = 0$, and so both sides are zero; otherwise

$$\frac{f(g(x)) - f(g(a))}{x - a} = \frac{f(g(x)) - f(g(a))}{g(x) - g(a)} \frac{g(x) - g(a)}{x - a}$$

and the result follows.

10.4   INTERNAL OBJECTS

10.4.1 It is now necessary to consider in more detail the sense in which $^*\mathbb{R}$ can really be said to have "the same properties" as $\mathbb{R}$. As remarked in section 10.1.3 it is certainly true that $^*\mathbb{R}$ inherits all the first order properties of $\mathbb{R}$. That is to say, every true statement about the members of $\mathbb{R}$ (the real numbers) transfers to a corresponding true statement about the members of $^*\mathbb{R}$ (the hyperreal numbers). For example the addition and multiplication of numbers in $^*\mathbb{R}$ are both commutative and associative, just as in $\mathbb{R}$: in fact, $^*\mathbb{R}$ is an ordered field, just as $\mathbb{R}$ is itself an ordered field. But, in contrast, consider once more the basic least upper bound property of $\mathbb{R}$, which states that every non-empty set of real numbers which is bounded above must have a least upper bound in $\mathbb{R}$. This is a property not of individual real numbers in $\mathbb{R}$ but of sets of real numbers. Therefore it is not a first order property, and it does not carry over to $^*\mathbb{R}$. Nevertheless there is still a sense in which $^*\mathbb{R}$ does enjoy a form of the least upper bound property, in that it applies to a certain distinguished class of subsets of $\mathbb{R}$. These are the so-called **internal** subsets of $^*\mathbb{R}$.

10.4.2 Internal sets. A subset Γ of $^*\mathbb{R}$ is said to be **internal** if there exists a sequence $(G_n)_{n \in \mathbb{N}}$ of subsets of $\mathbb{R}$ such that $x = [x_n]$ is a member of Γ if and only if $x_n \in G_n$ for nearly all n: it is then often convenient to write Γ explicitly as $[G_n]$. All other subsets of $^*\mathbb{R}$ are said to be **external**.

The nonstandard extension of a standard set G is a special case of an internal set in which $G_n = G$ for (nearly) all n. It is the internal subsets of $^*\mathbb{R}$ which inherit the properties of (appropriate) subsets of $\mathbb{R}$. In particular, the least upper bound property of $\mathbb{R}$ carries over to $^*\mathbb{R}$ in the following form:

Proposition 10.4. Every non-empty internal subset of $^*\mathbb{R}$ which is bounded above in $^*\mathbb{R}$ has a least upper bound in $^*\mathbb{R}$.

Proof. Let $A = [A_n]$ be a non-empty subset of $^*\mathbb{R}$ which is bounded above in $^*\mathbb{R}$. This will be true if and only if, for nearly all n, the standard set A_n is non-empty and bounded above in $\mathbb{R}$. In that case each such A_n will have a least upper bound a_n in $\mathbb{R}$, and then the hyperreal number $\alpha = [a_n]$ will be the required least upper bound of A in $^*\mathbb{R}$.

The fact that $\mathbb{R}$ is a subset of $^*\mathbb{R}$ which is bounded above in $^*\mathbb{R}$ but which has no least upper bound is no longer mysterious. It merely shows us that $\mathbb{R}$ is an external subset of $^*\mathbb{R}$.

Remark. Proposition 10.4 has corollaries of some importance.

Corollary 10.4. Let A be an internal subset of $^*\mathbb{R}$. Then,

(i) **Overflow:** If A contains arbitrarily large finite elements then A contains an infinite element.

(ii) **Underflow:** If A contains arbitrarily small positive infinite elements then A contains a finite element.

Proof. (i) If A is unbounded then there is nothing to prove. Otherwise let α be the least upper bound of A. Then α is surely infinite, and there exists $x \in A$ such that $\alpha/2 < x < \alpha$.

(ii) Let β be the greatest lower bound of the set A_+ of all positive elements of A. Then β is surely finite and there exists $y \in A$ such that $\beta < y < \beta + 1$.

10.4.3 Hypernatural numbers. If A is any (standard) subset of $\mathbb{R}$ then it is clear that its nonstandard extension $^*A$ is an internal set since it can be defined by the sequence $(A_n)_{n \in \mathbb{N}}$ in which $A_n = A$ for all $n \in \mathbb{N}$. In particular the nonstandard extension $^*\mathbb{N}$ of the set $\mathbb{N}$ of all natural numbers is internal. It inherits properties which are characteristic of $\mathbb{N}$, such as

(1) $^*\mathbb{N}$ is a discrete subset of $^*\mathbb{R}$ which contains all the natural numbers $1, 2, 3, \ldots$.

(2) $^*\mathbb{N}$ is closed under addition and multiplication.

(3) Each member Λ of $^*\mathbb{N}$ has an immediate successor $(\Lambda + 1)$ in $^*\mathbb{N}$.

(4) Each non-zero member Λ of $^*\mathbb{N}$ has an immediate predecessor $(\Lambda - 1)$ in $^*\mathbb{N}$.

(5) Every non-empty internal subset of $^*\mathbb{N}$ has a first element.

Once again it is important to note that in the case of (5) we are concerned with a property not of individual numbers but with sets of numbers. This is not a first order property of $\mathbb{N}$, and it does not carry over to $^*\mathbb{N}$, for it is easy to see that not every non-empty subset of $^*\mathbb{N}$ has a first element. In particular, the set $^*\mathbb{N}_\infty$ of all infinite hypernatural numbers is a non-empty subset of $^*\mathbb{N}$ which has no first element; hence it is another example of an external subset of $^*\mathbb{R}$.

10.4.4 Remarks.

(i) The existence of the set $^*\mathbb{N}$ of hypernatural numbers allows us to explain how the Archimedean property of $\mathbb{R}$ may be transformed into a meaningful and valid statement about *$\mathbb{R}$. As stated in Sec. 1.1.3 the Archimedean property in $\mathbb{R}$ may be expressed as follows:

If a and b are any two positive real numbers then there exists a positive integer (natural number), n, such that $a < nb$.

In *$\mathbb{R}$ there is a corresponding '*-Archimedean property' which we can state as

If α and β are any two positive hyperreal numbers then there exists a positive integer (hypernatural number), Λ, such that $\alpha < \Lambda\beta$.

No paradox results, precisely because the integer Λ may be infinite.

Hyperfinite sets. An internal subset A of *$\mathbb{R}$ is said to be **hyperfinite**, or ***-finite**, if there exists some hypernatural number $\Lambda \in$ *$\mathbb{N}$ and an internal one-to-one mapping F which carries the set $\{1, 2, \ldots, \Lambda\}$ into A. A necessary and sufficient condition for the internal set $A \equiv [A_n]$ to be hyperfinite is that the standard sets A_n are finite, for (nearly) all $n \in \mathbb{N}$.

The **internal cardinality** of the hyperfinite set $A \equiv [A_n]$ is the hypernatural number $| A | \equiv [|A_n|]$, where $|A_n|$ denotes the number of elements in the standard (finite) set A_n.

Examples.

(i) If A is any standard finite set then A is certainly internal and hyperfinite. Also the internal cardinality of A coincides with the number of elements of A in the usual, standard sense.

(ii) Let $\Lambda \equiv [m_n]$ be any infinite hypernatural number. Then the set

$$T_\Lambda = \{0, 1/\Lambda, 2/\Lambda, \ldots, (\Lambda - 1)/\Lambda, 1\}$$

is hyperfinite, with internal cardinality $\Lambda + 1$. For we have $T_\Lambda \equiv [T_n]$ where $T_n = \{0, 1/m_n, 2/m_n, \ldots, (m_n - 1)/m_n, 1\}$, and it follows at once that

$$|T_\Lambda| = [|T_n|] = [m_n + 1] = \Lambda + 1.$$

Hyperfinite sets inherit combinatorial properties of finite sets. Thus we have

(1) If A and B are hyperfinite sets then so also are $A \cup B$, $A \cap B$ and $A - B$.

(2) A hyperfinite union of hyperfinite sets is hyperfinite.

(3) If A is hyperfinite, and if $|A| = \Lambda$, then the nonstandard extension *$P(A)$ of the power set $P(A)$ is hyperfinite and $|{}^*P(A)| = 2^\Lambda$.

(4) Every non-empty hyperfinite set has a maximum and a minimum element.

(5) If α and β are any hyperreal numbers (with $\alpha \leq \beta$) then the set

$$\{\mu \in {}^*\mathbb{N} : \alpha \leq \mu \leq \beta\}$$

is hyperfinite.

10.4.5 Internal functions. Let $A = [A_n]$ be an internal subset of *$\mathbb{R}$ and suppose that there exists a sequence (f_n) of standard functions such that, for each n, f_n maps A_n into $\mathbb{R}$. Then the sequence (f_n) defines a function F mapping the internal set A into *$\mathbb{R}$ according to the rule

$$\text{for } x = [x_n] \in A,, \quad F(x) = [(f_n(x_n))_{n\in\mathbb{N}}]. \tag{10.9}$$

Any such function F is said to be an **internal function**. All other functions defined on *$\mathbb{R}$ or on some subset of *$\mathbb{R}$ are said to be external.

Once again, the nonstandard extension *f of a standard function f is a special case of an internal function in which $f_n = f$ for (nearly) all n.

Examples.

(i) For each $n \in \mathbb{N}$ the function

$$f_n(x) = \frac{n}{\pi(n^2x^2 + 1)}$$

is well defined on the whole of $\mathbb{R}$. The internal function $F = [f_n]$ is defined on the whole of *$\mathbb{R}$ and can be written as

$$F(x) = \frac{\Omega}{\pi(\Omega^2x^2 + 1)} \equiv \frac{\epsilon}{\pi(x^2 + \epsilon^2)}. \tag{10.10}$$

This is an internal function which is not the nonstandard extension of any standard function.

(ii) The function

$$H_{(t)}(x) = \frac{1}{2} + \frac{1}{\pi}\arctan(\Omega x)) \tag{10.11}$$

is again an internal function defined on the whole of $^*\mathbb{R}$ for which we have $H_{(t)} = [(h_n)_{n \in \mathbb{N}}]$ where the standard functions h_n are defined by

$$h_n(x) = \frac{1}{2} + \frac{1}{\pi}\arctan(nx).$$

This internal function is a nonstandard representative of the Heaviside unit step function: for every $x \in \mathbb{R}$ we have

$$H_{(t)}(x) \approx {}^*u_{1/2}(x)$$

where $u_{1/2}$ is the unit step function taking the value $1/2$ at the origin.

(iii) As an example of an external function we may take $E : {}^*\mathbb{R} \to {}^*\mathbb{R}$ where $E(x) = 1$ for all $x \approx 0$ and $E(x) = 0$ otherwise.

Hyperfinite sums. Given a hyperfinite set $A = [A_n]$ and an internal function $F = [f_n]$ defined (at least) on A, we can form the **hyperfinite sum of F over A** by setting

$$\sum_{a \in A} F(a) \equiv \left[\sum_{a_n \in A_n} f_n(a_n) \right]. \qquad (10.12)$$

Every elementary (Riemann) integral of a continuous function f can be represented as the standard part of a hyperfinite sum. Thus let f be defined and continuous on a finite real interval $[a, b]$. For each n in $\mathbb{N}$ let P_n be a partition of $[a, b]$ into n equal sub-intervals, each of length $d_n = (b - a)/n$. Then we have

$$\int_a^b f(x)dx = st. \left\{ \left[\sum_{k=1}^n f(a + kd_n)d_n \right] \right\}$$

$$\equiv st. \left\{ \sum_{x \in P_\Omega} {}^*f(a + x)d_\Omega \right\} \qquad (10.13)$$

where $P_\Omega = [P_n]$ is a hyperfinite partition of $^*[a, b]$, and $d_\Omega = [d_n]$.

10.4.6 Internal function calculus. Just as properties of subsets of $\mathbb{R}$ transfer to internal subsets of $^*\mathbb{R}$, so do properties of certain types of standard function transfer to corresponding types of nonstandard internal function. In particular, we can develop a calculus of internal functions which extends the ordinary processes of standard calculus. Corresponding to the concept of continuity for standard functions on $\mathbb{R}$ there is a property applicable to internal functions on $^*\mathbb{R}$ which we call *-**continuity**. Similarly we can define for internal functions a *-**derivative** and an elementary *-**integral**:

*-**continuity:** Let $F = [f_n]$, in which each f_n is a standard function which is continuous in the ordinary (standard) sense on $\mathbb{R}$. Given any hyperreal $\alpha \equiv [a_n] \in {}^*\mathbb{R}_{bd}$, let $\rho \equiv [r_n]$ be any positive hyperreal. Then for each n we can find $s_n \in \mathbb{R}_+$ such that $|f_n(x_n) - f_n(a_n)| < r_n$ for any x_n such that $|x_n - a_n| < s_n$. It follows that the internal function F is *-continuous at α in the following sense:

Given any positive hyperreal $\rho = [r_n]$ there exists a corresponding positive hyperreal $\sigma = [s_n]$ such that

$$|F(x) - F(\alpha)| < \rho$$

for all hyperreal x such that $|x - \alpha| < \sigma$.

Conversely, it can be shown that any internal function F which is *-continuous in this sense on $^*\mathbb{R}_{bd}$ must necessarily be of the form $F = [f_n]$ in which the f_n are continuous functions in the standard sense for nearly all n.

In standard analysis the set of all functions continuous everywhere on $\mathbb{R}$ is denoted by $C^0(\mathbb{R})$ or, more briefly, by $C(\mathbb{R})$. A necessary and sufficient condition for an internal function F to be *-continuous on $^*\mathbb{R}_{bd}$ is that $F = [f_n]$ where for (nearly) all $n \in \mathbb{N}$ each function f_n is continuous in the standard sense on $\mathbb{R}$. We shall denote the set of all such internal functions by $^*C(\mathbb{R})$. Clearly, a standard function f belongs to $C(\mathbb{R})$ if and only if its nonstandard extension $^*f$ belongs to $^*C(\mathbb{R})$. However, there exist functions in $^*C(\mathbb{R})$ which are not the nonstandard extensions of any standard function. Consider for example the function

$$H_{(t)}(x) = \frac{1}{2} + \frac{1}{\pi}\arctan(\Omega x).$$

This is *-continuous everywhere on $^*\mathbb{R}_{bd}$ but there is no standard function f, whether continuous or not, such that $^*f(x) = H_{(t)}(x)$ for all hyperreal x.

*-**derivative:** Let $F = [f_n]$ where each standard function f_n has a standard derivative $Df_n \equiv f_n'$. Then F is said to be *-differentiable

and we define the *-derivative of F to be the internal function given by

$$^*DF := [Df_n] \equiv [f'_n]. \tag{10.14}$$

As an example note that the internal function $F(x)$ defined in equation (10.10) above is the *-derivative of the internal function $H_{(t)}(x)$ defined in equation (10.11). It is clear that the definition of *-derivative agrees with that of the standard derivative whenever $F = {}^*f$ for some standard differentiable function f. The *-differential operator $^*D$ has the same formal properties as the standard differential operator D. Thus, it follows from the definition of $^*D$ that we have

(i) F, G *-differentiable $\Rightarrow (F + G)$ *-differentiable, and

$$^*D(F + G) = {}^*DF + {}^*DG,$$

(ii) F, G *-differentiable $\Rightarrow (FG)$ * -differentiable, and

$$^*D(FG) = ({}^*DF)G + F({}^*DG).$$

*-integral: Let $F = [f_n]$ where, for each n, the standard function f_n is (Riemann) integrable over the finite real interval $[a_n, b_n]$. If $\alpha = [a_n]$ and $\beta = [b_n]$ then we define the elementary *-integral of F over the internal interval $[\alpha, \beta]$ by

$$^*\int_\alpha^\beta F(x)dx \equiv \left[\left(\int_{a_n}^{b_n} f_n(x)dx \right)_{n \in \mathbb{N}} \right]. \tag{10.15}$$

If $[a_n, b_n] = [a, b]$ for all n, and if F is the nonstandard extension, $^*f$, of a function f integrable over $[a, b]$ then we simply get

$$^*\int_a^b {}^*f(x)dx = \left[\left(\int_a^b f(x)dx \right) \right] = \int_a^b f(x)dx$$

so that in this case the nonstandard integral coincides with the standard one. In general, properties of the standard integral carry over to the nonstandard one. For example, we have:

Proposition 10.5 (*Mean Value Theorem). Suppose that the internal function F is *-continuous on the *-bounded hyperreal interval $[\alpha, \beta]$. Then there exists some hyperreal number τ in $[\alpha, \beta]$ such that

$$^*\int_\alpha^\beta F(x)dx = F(\tau)(\beta - \alpha). \tag{10.16}$$

Proof. Let $F = [f_n]$ where each f_n is continuous on an interval $[a_n, b_n]$. Then by the standard Integral Mean Value Theorem there exists, for each n, a real $t_n \in [a_n, b_n]$ such that

$$\int_{a_n}^{b_n} f_n(x)dx = f_n(t_n)(b_n - a_n)$$

Hence,

$$* \int_\alpha^\beta F(x)dx = \left[\int_{a_n}^{b_n} f_n(x)dx \right] = [f_n(t_n)(b_n - a_n)]$$

$$= F(\tau)(\beta - \alpha), \quad \text{where} \quad \tau = [t_n],$$

which establishes (10.16).

Similarly, it is easy to confirm that the generalised Mean Value theorem of the standard integral calculus has an appropriate analogue for the nonstandard integral:

If $F(t)$ is *-continuous and $G(t)$ is positive on $[\alpha, \beta]$ then there exists a hyperreal τ in $[\alpha, \beta]$ such that

$$* \int_\alpha^\beta F(t)G(t)dt = F(\tau) * \int_\alpha^\beta G(t)dt. \tag{10.17}$$

A version of the partial integration formula is also valid for the nonstandard integral:

Proposition 10.6. If F and G are *-differentiable internal functions such that $*DF$ and $*DG$ are *-continuous on $[\alpha, \beta] \subset *\mathbb{R}_{bd}$, then we have

$$* \int_\alpha^\beta F(x) *DG(x)dx = F(\beta)G(\beta) - F(\alpha)G(\alpha)$$

$$-* \int_\alpha^\beta *DF(x)G(x)dx \tag{10.18}$$

The proof of Proposition (10.6) is left as an exercise for the reader.

Finally, note that although the definition of this nonstandard elementary integral has been given wholly in terms of standard Riemann integrals taken over standard finite intervals, the nonstandard integral may well turn out to be defined for a hyperreal interval which may not even be finite. It is easy to see that improper Riemann integrals can readily be defined in terms of nonstandard integrals taken over such intervals as $[\varepsilon, \Omega]$, $[-\Omega, -\varepsilon]$ and $[\Omega, +\Omega]$; moreover even an integral over $[-\varepsilon, +\varepsilon]$ can make sense. This provides appropriate and significant interpretations of such otherwise conceptually confusing conventions as,

$$\int_{0-}^{+\infty} f(x)dx, \quad \int_{0+}^{+\infty} f(x)dx, \quad \text{and} \quad \int_{0-}^{0+} f(x)dx.$$

Further reading.

A number of introductory accounts of Nonstandard Analysis are currently available. The original book **Nonstandard Analysis** by **Abraham Robinson** has appeared in a revised edition published in 1996 by Princeton University Press, and is still the ultimate source for the essential theory. However, as remarked earlier, this is not perhaps the easiest introduction to the subject but it is a classic text and it contains an astonishingly comprehensive presentation of the application of nonstandard methods. In complete contrast the **Elementary Calculus: An Infinitesimal Approach** by **H.Jerome Keisler**, (Prindle, Weber and Schmidt, 1986) uses an extremely simple basic approach to NSA to give a thoroughgoing first course in calculus from a nonstandard viewpoint, assuming little more than a basic acquaintance with elementary algebra. A somewhat more demanding, but still readily approachable, introduction to real analysis is provided by the book **Infinitesimal Calculus** by **J.M.Henle** and **I.Kleinberg**, (MIT Press, 1979). Another, experimental, attempt to explain the principles of nonstandard methods at a relatively elementary level has been made in **Standard and Nonstandard Analysis** by R.F. Hoskins,(Ellis Horwood, 1990). This is an otherwise orthodox introduction to real analysis in which nonstandard techniques using infinitesimals are presented side by side with standard arguments.

But perhaps the most readable introduction to the fundamental principles of NSA itself is that given by **Tom Lindstrom** in the LMS Student Text publication **Nonstandard Analysis and its Applications** edited by **Nigel Cutland**, C.U.P., 1988): the content of Chapter 10 above is largely based on Lindstrom's account. This LMS text also contains some more recent applications of NSA to probability theory, functional analysis and mathematical physics. More advanced material on the theoretical foundations of nonstandard analysis, together with applications to general topology, functional analysis and the theory of measure and integration will be found in **Introduction to the Theory of Infinitesimals** by **K.D.Stroyan** and **W.A.J.Luxemburg**, (Academic Press, 1976), in **An Introduction to Nonstandard Real Analysis** by **A.E.Hurd** and **P.A.Loeb**, (Academic Press, 1985), and more recently in **Lectures on the Hyperreals** by **Robert Goldblatt**, (Springer, 1998). None of these, however, are primarily concerned with the possibilities of an extended nonstandard treatment of generalised functions.

One other text, of a somewhat different character, deserves particular mention. **Infinitesimal Methods of Mathematical Analysis** by **J.Sousa Pinto**, (Horwood Publishing, 2004 begins with a careful and informative account of the non-archimedian fields studied by a number of mathematicians before the advent of Nonstandard Analysis proper in the 1960s. This leads to a simple introductory account of NSA with specific applications to generalised functions which is of the same character as that explored in the following chapter of the present text, and is accordingly strongly recommended as supplementary reading.

Chapter 11

Nonstandard Generalised Functions

11.1   NONSTANDARD δ-FUNCTIONS

11.1.1 An elementary nonstandard delta function. Armed with the knowledge that we may now call on the existence of a nonstandard extension of the real number system $\mathbb{R}$, we return here to the problem posed at the beginning of Chapter 2: that is, to obtain a satisfactory definition of a function $\delta(t)$ which would fulfill the requirements made explicit by Paul Dirac. Following Lützen we can list what seems to be required as the essential chararacteristic properties of the Dirac delta function as:

(DF1) $\delta(t) = \frac{d}{dt}u(t)$;
(DF2) $\delta(t) = \lim_{n \to \infty} d_n(t)$, for suitably chosen functions $d_n(t)$;
(DF3) $\delta(t) = 0$ for $t \neq 0$, and $\int_{-\infty}^{+\infty} \delta(t)dt = 1$;

(DF4) $\int_{-\infty}^{+\infty} f(t)\delta(t-a)dt = f(a)$, at least for all functions $f(t)$ sufficiently smooth at $t = a$.

In section 2.3.2 an elementary 'standard' interpretation of the Dirac delta function, based on (DF2), was given in terms of sequences of very simple ordinary functions:

$$s_n(t) = \begin{cases} 1 & \text{for } t > 1/2n \\ nt + 1/2 & \text{for } -1/2n \leq t \leq +1/2n \\ 0 & \text{for } t < -1/2n \end{cases}$$

237

$$d_n(t) = \begin{cases} 0 & \text{for } t > 1/2n \\ n & \text{for } -1/2n < t < +1/2n \\ 0 & \text{for } t < -1/2n \end{cases}$$

and the sampling property (DF4) attributed to $\delta(t)$ was expressed in terms of a limit,

$$\int_{-\infty}^{+\infty} f(t)\delta(t)dt \equiv \lim_{n\to\infty} \int_{-\infty}^{+\infty} f(t)d_n(t)dt = f(0). \qquad (11.1)$$

It was made clear that the symbolic expression $\delta(t)$ could not be identified with the pointwise limit $d_\infty(t)$ of the sequence $(d_n(t))_{n\in N}$ since for every continuous function $f(t)$ we must have

$$\int_{-\infty}^{+\infty} f(t)d_\infty(t)dt = 0. \qquad (11.2)$$

This admission of the purely symbolic role played by expressions like

$$\int_{-\infty}^{+\infty} f(t)\delta(t)dt$$

is unavoidable in the context of standard analysis, and the conception of $\delta(t)$ as a function defined on $\mathbb{R}$ must be acknowledged as only a convenient fiction there. Interpretation of both (DF1) and (DF3) was therefore not immediately obvious.

Suppose now that, instead of working in terms of the standard real number system $\mathbb{R}$, we have to hand a simple model of a hyperreal number system $^*\mathbb{R}$ together with definitions of internal sets and functions as described in Chapter 10 above. The sequences $(s_n)_{n\in N}$ and $(d_n)_{n\in N}$ generate internal functions on $^*\mathbb{R}$ which we will denote by $\eta_{(d)}$ and $\delta_{(d)}$ respectively:

$$\eta_{(d)}(t) = \begin{cases} 1 & \text{for } t \geq 1/2\Omega \\ \Omega t + 1/2 & \text{for } |t| < 1/2\Omega \\ 0 & \text{for } t \leq 1/2\Omega \end{cases}$$

and

$$\delta_{(d)}(t) = \begin{cases} 0 & \text{for } t \geq 1/2\Omega \\ \Omega & \text{for } |t| < 1/2\Omega \\ 0 & \text{for } t \leq -1/2\Omega \end{cases}$$

where, as before, we use the special symbol Ω for the infinite hyperreal $[n]$.

It is clear that $\eta_{(d)}(t)$ coincides with the Heaviside unit step function $u(t)$ on $\mathbb{R}$, and that $\delta(t)$ is the *-derivative of $\eta(t)$. Now let f be any standard function which is continuous at least on some neighbourhood of the origin. Then the product $^*f(t)\delta_{(d)}(t)$ will be *-integrable over any hyperreal interval $(-\Lambda, \Lambda)$, where Λ is any positive infinite hyperreal. Further, the First Mean Value Theorem of the integral calculus applies to *-integrals and so we get

$$^*\int_{-\Lambda}^{\Lambda} {}^*f(t)\delta_{(d)}(t)dt = {}^*\int_{-1/2\Omega}^{1/2\Omega} {}^*f(t)\Omega dt = {}^*f(\xi),$$

where ξ is some point in $(-1/2\Omega, 1/2\Omega)$. By the continuity of f at the origin we have $^*f(\xi) \approx f(0)$. Hence, since Λ may be any infinite positive hyperreal whatsoever, it would make sense to write

$$\int_{-\infty}^{+\infty} f(t)\delta(t)dt \equiv st. \left\{ {}^*\int_{-\Lambda}^{\Lambda} {}^*f(t)\delta_{(d)}(t)dt \right\} = f(0), \qquad (11.3)$$

in order to suggest another possible interpretation for the symbolic integral appearing on the left-hand side, and for the Dirac symbol δ.

The immediate advantage of such an interpretation is that the sampling property (DF4) attributed formally to the symbol δ can now be legitimately associated with a genuine function, $\delta_{(d)}$, albeit one which maps *$\mathbb{R}$ into *$\mathbb{R}$. We shall call such an internal function $\delta_{(d)}$ a **pre-delta function**. Note also that there is no longer any need to formulate a Stieltjes integral representation in this context, and little to be gained in doing so, since for any $n \in \mathbb{N}$ we surely have

$$\int_{-1/2n}^{+1/2n} f(t)ds_n(t) = \int_{-1/2n}^{+1/2n} f(t)d_n(t)dt.$$

Hence, for any positive infinite hyperreals Γ, Λ, it is easy to see that

$$^*\int_{-\Lambda}^{+\Lambda} {}^*f(t)d\eta_{(d)}(t) \approx {}^*\int_{-\Gamma}^{+\Gamma} {}^*f(t)\delta_{(d)}(t)dt.$$

11.1.2 Comparison with pointwise limit $^*d_\infty(t)$. The internal function $\delta_{(d)}$ vanishes outside the monad of 0 and hence is zero for all non-zero real t. Although it therefore coincides with the pointwise limit $d_\infty(t)$ on $\mathbb{R} - \{0\}$, there is now no conflict between (11.3) and (the nonstandard equivalent of) (11.2). The nonstandard extension $^*d_\infty(t)$ of the pointwise limit $d_\infty(t)$ of the sequence $(d_n)_{n\in\mathbb{N}}$ is quite clearly a different function from $\delta_{(d)}(t)$. It is well-defined as zero for every non-zero hyperreal value of t, and its $*$-integral over any hyperreal integral is zero. Therefore there can be no sampling property associated with $^*d_\infty(t)$.

In fact we can make this point even more strongly by adopting a slightly different approach to the definition of the internal function acting as a representative of the delta function. For $n = 1, 2, \ldots$ we have

$$d_n(t - 1/2n) = \begin{cases} 0 & \text{for } t \geq 1/n \\ n & \text{for } 0 < t < 1/n \\ 0 & \text{for } x \leq 0 \end{cases}$$

Using these functions instead of the $d_n(t)$ defines an internal function $\delta_{(d)}(t-1/2\Omega)$ which has the value Ω for all hyperreal t such that $0 < t < \Omega$ and vanishes everywhere else on $^*\mathbb{R}$. This function exhibits the same kind of sampling property (11.3) as $\delta_{(d)}(t)$. But this time the pointwise limit of the standard sequence $(d_n(t - 1/2n))_{n\in\mathbb{N}}$ is defined everywhere on $\mathbb{R}$ as the function which is identically zero, and so its nonstandard extension vanishes identically on $^*\mathbb{R}$. On the other hand the internal function $\delta_{(d)}(t-1/2\Omega)$ is a nonstandard representative of the Dirac delta function which differs from zero only at those (nonstandard) points lying between 0 and $1/\Omega$: the 'spike' (i.e. the business end of the delta function) now appears to be a purely hyperreal phenomenon, with no "real" existence.

11.1.3 A nonstandard theory of the Dirac delta function, such as that described above, is both rigorous and relatively simple to understand. Conventional acccounts such as are often offered to physicists, engineers and other users of mathematics whose interests are considered to be with practical applications are almost inevitably non-rigorous and yet fail to achieve conceptual clarity. The most immediate advantage of a nonstandard approach is the ability to use infinitesimals and infinite numbers freely and thereby treat the delta function as a genuine function. But this is not all. Algebraic operations involving $\delta_{(d)}$ can be carried out in an entirely straightforward manner, without the special qualifications which seem to be needed when dealing with the conventional symbol δ. Much of the painstaking explanation of the algebra of delta functions elaborated in Chapter 3 becomes redundant. Thus, the sum of $\delta_{(d)}(t)$ and an 'ordinary' function $h(t)$ (or, strictly, its nonstandard extension $^*h(t)$) is defined by pointwise addition in the obvious way, as also is the sum of any translates $\delta_{(d)}(t-a)$, $\delta_{(d)}(t-b)$ etc. Similarly the behaviour of products involving $\delta_{(d)}(t)$ is easy to understand. If $\phi(t)$ and $f(t)$ are each continuous on $\mathbb{R}$ then for any positive infinite hyperreal Λ we have

$$^* \int_{-\Lambda}^{+\Lambda} {}^*f(t)\{{}^*\phi(t)\delta_{(d)}(t)\}dt = {}^* \int_{-\Lambda}^{+\Lambda} \{{}^*f(t)\,{}^*\phi(t)\}\delta_{(d)}(t)dt$$

$$= {}^*f(\xi)\,{}^*\phi(\xi) \approx kf(0)$$

where $k = \phi(0)$. Hence the internal function $^*\phi(t)\delta_{(d)}(t)$ behaves as a nonstandard representative of a delta function of weight k, even though we do not have $^*\phi(t)\delta_{(d)}(t) = k\delta_{(d)}(t)$ in general. Once again there is an obvious contrast with the behaviour of the pointwise limit, $d_\infty(t)$: if, for example, k is positive then we have $kd_\infty(t) = d_\infty(t)$, and the action of $kd_\infty(t)$ on an arbitrary continuous function $f(t)$ remains equivalent to that of the null function.

The fact that the product $u\delta$ seems to be indeterminate (Sec.3.2.3) should cause no surprise once a nonstandard setting is examined. The internal function $^*u(t)\delta_{(d)}(t)$ is well-defined on $^*\mathbb{R}$ by

$$^*u(t)\delta_{(d)}(t) = \begin{cases} 0 & \text{for } t \geq 1/2\Omega \\ \Omega & \text{for } 0 < t < 1/2\Omega \\ 0 & \text{for } t \leq 0 \end{cases}$$

and is easily seen to behave as a nonstandard representative of $\frac{1}{2}\delta(t)$. On the other hand if we consider the product $^*u(t)\delta_{(d)}(t - 1/2\Omega)$ then this is just as easily seen to act as a nonstandard representative of $1\delta(t) \equiv \delta(t)$. Finally, replacing $\delta_{(d)}(t - 1/2\Omega)$ by $\delta_{(d)}(t + 1/2\Omega)$, we would get $^*u(t)\delta_{(d)}(t + 1/2\Omega) \equiv 0$, and this can only represent the formal product $0\delta(t) = 0$.

Nevertheless it may be argued, and indeed has been, that taking on board the necessary background of nonstandard analysis is too high a price to pay. Other, simpler, examples of field extensions of $\mathbb{R}$ containing bona fide infinitesimal and infinite elements are known to exist and could be exploited to provide conceptually attractive models of delta functions. A particularly interesting attempt to do this has been developed by **B.D.Craven** in a paper **Generalized functions for applications**, (J. Austral. Math. Soc. Ser. B, 1985, 362-374). However, this lacks the power which nonstandard analysis provides and results in a greatly restricted theory of generalised functions. Criticism of N.S.A. on the grounds of difficulty are ill-founded. As the account given in Chapter 10 should show, there are no serious obstacles to understanding the basic principles and techniques. And the advantages of applying nonstandard methods to the theory of generalised functions are much more than those conceptual gains already referred to. In the next two sections this is illustrated by a more extended study of nonstandard delta functions, following the pioneer work of **Detlef Laugwitz**.

11.2   PRE-DELTA FUNCTIONS

11.2.1 Pre-delta functions with kernel. There are, in fact, many different kinds of internal function on $^*\mathbb{R}$, each exhibiting some form of the delta function sampling property. We will generally refer to any such function as a pre-delta function. Thus let $\theta(t)$ be any standard function on $\mathbb{R}$ which is bounded on $\mathbb{R}$ and which is such that $\int_{-\infty}^{+\infty} \theta(t)dt = 1$. If, for each $n \in \mathbb{N}$, we write $\theta_n(t) := n\theta(nt)$ then the sequence $(\theta_n)_{n \in \mathbb{N}}$ generates an internal function on $^*\mathbb{R}$ which we shall denote by $\delta_{(\theta)}$. For any $t = [t_n]$ in $^*\mathbb{R}$ we have

$$\delta_{(\theta)}(t) = [\theta_n(t_n)] \equiv [n\theta(nt_n)]$$

or, equivalently,

$$\delta_{(\theta)}(t) = \Omega\, ^*\theta(\Omega t), \tag{11.4}$$

where, as before, we use the symbol Ω to denote the infinite hyperreal number defined by the equivalence class $[n]$. We shall refer to such an internal function as a **pre-delta function with kernel** θ.

If, in particular, $\theta(t)$ vanishes identically outside some finite real interval $(-a, a)$ then the internal function $\delta_{(\theta)}$ exhibits the kind of pointwise behaviour popularly attributed to the Dirac delta function in the following sense: $\delta_{(\theta)}(t)$ vanishes identically outside the infinitesimal hyperreal interval

$$(-a/\Omega, a/\Omega) \subset \text{monad}(0)$$

and therefore in particular $\delta_{(\theta)}(t) = 0$ for all real, (i.e. standard) non-zero t. Also $\delta_{(\theta)}(t)$ will assume infinite values at some points within the infinitesimal, hyperreal interval $(-a/\Omega, a/\Omega)$, although not necessarily at the point $t = 0$ itself. (In fact if $\theta(0) = 0$ then $\delta_{(\theta)}(t)$ must vanish for all real values of t.)

However, if $\theta(t)$ does not vanish outside some finite real interval, then $\delta_{(\theta)}(t)$ need not show such supposedly "characteristic" Dirac delta function behaviour. The fact that $\delta_{(\theta)}$ is "concentrated at the origin" remains true, but only in the following, less direct, sense:

Let $\Lambda = [(m_n)_{n \in \mathbb{N}}]$ be any positive infinite integer. Then the nonstandard integral of $*\theta(t)$ is well defined as

$$* \int_{-\Lambda}^{\Lambda} *\theta(t)dt = \left[\left(\int_{-m_n}^{m_n} \theta(t)dt \right)_{n \in \mathbb{N}} \right] \approx 1, \qquad (11.5)$$

and it follows readily that

$$1 \approx * \int_{-\Lambda}^{\Lambda} *\theta(t)dt = * \int_{-\Lambda/\Omega}^{\Lambda/\Omega} \theta(\Omega t)d(\Omega t)$$

$$= * \int_{-\Lambda/\Omega}^{\Lambda/\Omega} \delta_{(\theta)}(\tau)d\tau \qquad (11.6)$$

and that

$$* \int_{-\alpha}^{-\Lambda/\Omega} \delta_{(\theta)}(\tau)d\tau \approx 0 \approx * \int_{\Lambda/\Omega}^{\alpha} \delta_{(\theta)}(\tau)d\tau \qquad (11.7)$$

where α is any (positive) hyperreal such that $\alpha > \Lambda/\Omega$. Note that we may take Λ here to be any positive infinite hyperreal, not necessarily an integer and not necessarily less than or equal to Ω. Taking $\Lambda = \sqrt{\Omega}$, for example, gives $\Lambda/\Omega \approx 0$, which shows that all the effective "weight" of the pre-delta function is indeed confined to the monad of 0.

11.2.2 Non-negative kernel. Suppose first that $\theta(t)$ is a non-negative function everywhere on $\mathbb{R}$. It follows at once that the internal function $\delta_{(\theta)}(t)$ will then be non-negative everywhere on $*\mathbb{R}$. In this case the sampling property associated with the pre-delta function δ_θ can be derived by a mild extension of the argument used for $\delta_{(d)}$:

If f is bounded and continuous on $\mathbb{R}$ then, for any positive infinitesimal ξ, and any positive infinite hyperreal Λ, we have

$$^*\!\int_{-\Lambda}^{\Lambda} {}^*\!f(t)\delta_{(\theta)}(t)dt = {}^*\!\int_{-\Lambda}^{\xi} {}^*\!f(t)\delta_{(\theta)}(t)dt$$
$$+ {}^*\!\int_{\xi}^{\Lambda} {}^*\!f(t)\delta_{(\theta)}(t)dt + {}^*\!\int_{-\xi}^{\xi} {}^*\!f(t)\delta_{(\theta)}(t)dt.$$

Since f is bounded and continuous, it follows from (11.7) that the first two integrals on the right-hand side must be infinitesimal in value. Also, applying the (generalised) mean value theorem of the integral calculus to the third integral, we have

$$^*\!\int_{-\xi}^{\xi} {}^*\!f(t)\delta_{(\theta)}(t)dt \approx {}^*\!f(\tau) {}^*\!\int_{-\xi}^{\xi} \delta_{(\theta)}(t)dt \approx f(\tau)$$

where τ is some point in the infinitesimal hyperreal interval $(-\xi, \xi)$. By continuity we have $^*\!f(\tau) \approx f(0)$ and so we could write,

$$\int_{-\infty}^{+\infty} f(t)\delta(t)dt \equiv st\left\{ {}^*\!\int_{-\Lambda}^{\Lambda} {}^*\!f(t)\delta_{(\theta)}(t)dt \right\} = f(0) \qquad (11.8)$$

for any positive infinite hyperreal Λ.

Examples of such pre-delta functions are $\delta_{(d)}$, as already discussed, and

$$\Omega/\pi(\Omega^2 t^2 + 1) \quad ; \quad \sin^2(\Omega t)/\pi\Omega t^2 \quad ; \quad \frac{\Omega}{\sqrt{\pi}}\exp(-\Omega^2 t^2).$$

The result of (11.8) continues to hold if we drop the assumption that $\theta(x)$ is everywhere non-negative, and merely require that it is an absolutely integrable function over $(-\infty, +\infty)$. For then $\theta(x)$ can always be expressed as the difference of two non-negative functions, each integrable over $(-\infty, +\infty)$, and the argument given above requires only minor modification. Hence we have the result

Proposition 11.1. Let $\theta(t)$ be a standard function, defined and bounded on $\mathbb{R}$, which is absolutely integrable over $(-\infty, +\infty)$. If $\int_{-\infty}^{+\infty} \theta(t)dt = 1$ then, for any bounded continuous function $f(t)$ and any positive infinite hyperreal Λ, we have

$$^*\!\int_{-\Lambda}^{\Lambda} {}^*\!f(t)\delta_{(\theta)}(t) \approx f(0). \qquad (11.9)$$

11.2.3 At this stage it is tempting to suggest that the problem of interpreting the delta function symbolism may be resolved, once for all, by defining a suitable equivalence relation for internal functions on $^*\mathbb{R}$ and letting δ stand for the equivalence class containing all pre-delta functions. Such an equivalence relation could take the following form:

$F_1(t) \sim F_2(t)$ if and only if for every sufficiently well-behaved standard function f, we have

$$^*\!\int_{-\infty}^{+\infty} {}^*\!f(t)F_1(t)dt \approx {}^*\!\int_{-\infty}^{+\infty} {}^*\!f(t)F_2(t)dt. \tag{11.10}$$

Taking sufficiently well-behaved to mean bounded and continuous would allow us to identify the Dirac δ with an equivalence class $[\delta_{(\theta)}]$ containing all pre-delta functions generated by absolutely integrable kernel functions. This corresponds very closely with the view taken in the formal treatment of the Dirac delta function developed in the early chapters of this book, and accordingly we shall now refer to all pre-delta functions satisfying the conditions of Proposition 11.1 as **pre-delta functions of Dirac type**. However there exist important examples of pre-delta functions generated by kernels which are not absolutely integrable over $(-\infty, +\infty)$, and for which the statement of Proposition (11.1) needs to be modified.

11.2.4 General kernel. Let θ be a pre-delta kernel function which is not necessarily everywhere non-negative, and for which the integral $\int_{-\infty}^{+\infty} \theta(t)dt$ may converge to 1 only conditionally. Further, let $\eta_{(\Lambda,\theta)}$ denote the internal function defined by

$$\eta_{(\Lambda,\theta)}(t) := {}^*\!\int_{-\Lambda}^{t} \delta_{(\theta)}(\tau)d\tau \approx \begin{cases} 0 & \text{for } t \leq -1/\sqrt{\Omega} \\ 1 & \text{for } t \geq 1/\sqrt{\Omega} \end{cases}$$

with $\eta_{(\Lambda,\theta)}(t)$ set equal to 0 for all $t \leq -\Lambda$. If the kernel function were everywhere non-negative then we would have had

$$|\eta_{(\Lambda,\theta)}(t)| \leq 1$$

everywhere on $^*\mathbb{R}$. In general, however, this need not be true.

Suppose instead that we merely assume that $\eta_{(\Lambda,\theta)}(t)$ is always finite-valued on $^*\mathbb{R}$. This will certainly be the case if $\theta(t)$ is continuous on $\mathbb{R}$

(or, at least, has no worse than finitely many jump discontinuities). For then we can write

$$\eta_{(\Lambda,\theta)}(t) = \eta_{(\Lambda,\theta)}(0) + {}^*\!\int_0^{\Omega t} {}^*\theta(\tau)d\tau,$$

and since the standard integral $\int_{-\infty}^{+\infty} \theta(t)dt$ is convergent it follows that $|{}^*\!\int_a^{\Omega t} {}^*\theta(t)dt| \le 1$ for some finite real a, and therefore that $\eta_{(\Lambda,\theta)}$ is bounded on ${}^*\mathbb{R}$. That is to say, even if the non-negativity condition is not fulfilled we still have the (weaker) property

(B) There exists some real positive M such that $|\eta_{(\Lambda,\theta)}| \le M$ on ${}^*\mathbb{R}$.

This condition allows us to establish a sampling property for $\delta_{(\theta)}$ which, however, applies to a more restricted family of test functions:

Proposition 11.2. Let $\delta_{(\theta)}(t) = \Omega {}^*\theta(\Omega t)$ where $\theta(t)$ is continuous (or has only finitely many jump discontinuities) and is such that $\int_{-\infty}^{+\infty} \theta(t)dt = 1$. Then, if f is any standard function which is continuously differentiable on $\mathbb{R}$ and which vanishes outside some finite interval, say $(-a, a)$, we have

$$st\left\{ {}^*\!\int_{-\Lambda}^{\Lambda} {}^*f(t)\delta_{(\theta)}(t)dt \right\} = f(0).$$

for any positive infinite hyperreal Λ.

Proof. The proof is a rigorous version of the formal appeal to the integration by parts formula used in Sec. 2.2.2. Integration by parts gives

$$ {}^*\!\int_{-\Lambda}^{t} {}^*f(\tau)\delta_{(\theta)}(\tau)d\tau \approx {}^*\!\int_{-a}^{t} {}^*f(\tau)\delta_{(\theta)}(\tau)d\tau$$

$$= \eta_{(\Lambda,\theta)}(t)\,{}^*f(t) - {}^*\!\int_{-a}^{t} \eta_{(\Lambda,\theta)}(\tau)\,{}^*f'(\tau)d\tau.$$

Now since $\eta_{(\Lambda,\theta)}(t) \approx 0$ for $t \le -1/\sqrt{\Omega}$ it follows first that

$$ {}^*\!\int_{-a}^{-1/\sqrt{\Omega}} \eta_{(\Lambda,\theta)}(\tau)\,{}^*f'(\tau)d\tau \approx 0.$$

Next, using the fact that the kernel satisfies condition (B), we also have

$$\left| {}^*\!\int_{-1/\sqrt{\Omega}}^{1/\sqrt{\Omega}} \eta_{(\Lambda,\theta)}(\tau)\,{}^*f'(\tau)d\tau \right| \le 2MK/\sqrt{\Omega} \approx 0,$$

where we write $K = \sup_{|\tau| \le 1/\sqrt{\Omega}} \{|f'(\tau)|\}$. Hence,

$$* \int_{-\Lambda}^{t} {}^{*}f(\tau)\delta_{(\theta)}d\tau \approx \eta_{(\Lambda,\theta)} \, {}^{*}f(t) - {}^{*} \int_{1/\sqrt{\Omega}}^{t} \eta_{(\Lambda,\theta)}(\tau) \, {}^{*}f(\tau)d\tau.$$

Then, for $t \ge 1/\sqrt{\Omega}$, we have $\eta_{(\Lambda,\theta)} \approx 1$, and so

$$* \int_{1/\sqrt{\Omega}}^{t} \eta_{(\Lambda,\theta)}(\tau) \, {}^{*}f'(\tau)d\tau - {}^{*} \int_{1/\sqrt{\Omega}}^{t} f'(\tau)d\tau$$

$$= {}^{*} \int_{1/\sqrt{\Omega}}^{t} \{\eta_{(\Lambda,\theta)}(\tau) - 1\} \, {}^{*}f'(\tau)d\tau \approx 0,$$

since $f'(t)$ must vanish for $t \ge a$. Hence,

$$* \int_{-\Lambda}^{t} {}^{*}f(\tau)\delta_{(\theta)}(\tau)d\tau \approx {}^{*}f(t) - {}^{*} \int_{-1/\sqrt{\Omega}}^{t} {}^{*}f'(\tau)d\tau$$

$$\approx {}^{*}f(1/\sqrt{\Omega}) \approx f(0) \ , \quad \text{for } t \ge 1/\sqrt{\Omega}.$$

Thus for a kernel function θ which is continuous, but not absolutely integrable over $(-\infty, +\infty)$, there will still be a sampling property for $\delta_{(\theta)}$, at least with respect to standard functions which are continuously differentiable on $\mathbb{R}$ and which vanish outside some finite real interval. An important example of such a pre-delta function is obtained if we take for kernel the function $\theta(t) = \sin(t)/\pi t$. This is continuous on $\mathbb{R}$, and the integral $\int_{-\infty}^{+\infty} \theta(t)dt$ is conditionally convergent to 1. Hence conditions for Proposition 11.2 are fulfilled and the corresponding pre-delta function,

$$\delta_{(\theta)}(t) \equiv \sin(\Omega t)/\pi t \tag{11.11}$$

has a sampling property valid at least for all continuously differentiable functions of compact support. In point of fact the sampling property for this pre-delta function holds for a much wider class of functions than this, though not for all continuous functions. (There are classical examples of continuous functions for which the limit

$$\lim_{m \to \infty} \int_{-\infty}^{+\infty} f(t)\{\sin(mt)/\pi t\}dt$$

does not exist.)

11.3 PERIODIC DELTA FUNCTIONS

11.3.1 The recognition of different types of delta function is particularly valuable in Fourier analysis. Consider to begin with the simplest example of a periodic impulse train. As before, for $n = 1, 2, \ldots$, we write $d_n(t)$ for the standard function which takes the value n for all t such that $|t| < 1/2n$ and is equal to zero elsewhere on $\mathbb{R}$. The periodic extension of this function, of period 2π, is the train of rectangular pulses given by the infinite series

$$\sum_{m=-\infty}^{+\infty} d_n(t - 2\pi m). \tag{11.12}.$$

In a naive standard approach (as in Chapter 6) the periodic function, $\sum_{m=-\infty}^{+\infty} d_n(t-2\pi m)$ is said to 'converge', as $n \to \infty$, to an infinite periodic train of delta functions,

$$\sum_{m=-\infty}^{+\infty} \delta(t - 2\pi m).$$

What is actually meant is that given any continuous function f which vanishes outside some finite interval (a, b), there will be integers m_1, m_2 for which we will have (as $n \to \infty$),

$$\int_{-\infty}^{+\infty} \left\{ \sum_{m=-\infty}^{+\infty} d_n(t - 2\pi m) \right\} f(t)dt \longrightarrow \sum_{m=m_1}^{m=m_2} f(2\pi m).$$

That is to say, the characteristic sampling property of the (Dirac) delta function is manifest at the points $t = 2\pi m$, $(m = 0, \pm 1, \pm 2, \ldots)$.

Using a nonstandard formulation we would say instead that the sequence $\left(\sum_{m=-n}^{n} d_n(t - 2\pi m) \right)_{n \in \mathbb{N}}$ generates the internal function

$$\Delta_{(2\pi, d)}(t) = \left[\left(\sum_{m=-n}^{n} d_n(t - 2\pi m) \right)_{n \in \mathbb{N}} \right]$$

$$\equiv \sum_{m=-\Omega}^{\Omega} \delta_{(d)}(t - 2\pi m) \tag{11.13}$$

which is a periodic sum of (simple) Dirac pre-delta functions. If $f(t)$ is continuous and vanishes outside the finite interval (a, b) then

$$\int_{-\infty}^{+\infty} \Delta_{(2\pi, d)}(t) * f(t) dt \approx \sum_{m=m_1}^{m=m_2} f(2\pi m). \qquad (11.14)$$

If we now ask for a "Fourier series" representation of the periodic delta function train, then the situation in the standard approach is less clear. A Fourier series expansion of the periodic function (11.12) yields the result that, for all t such that $-\pi < t < \pi$,

$$\sum_{m=-\infty}^{+\infty} d_n(t - 2\pi m) = \frac{1}{2\pi} \sum_{m=-\infty}^{+\infty} c_m e^{imt}$$

$$= \frac{1}{2\pi} \left\{ 1 + \sum_{m=-\infty}^{+\infty} c_m \cos(mx) \right\}$$

where $c_m = \sin(m/2n)/(m/2n)$. As $n \to \infty$ then for each value of $m \in \mathbb{Z}$ the Fourier coefficient c_m tends to unity. Hence the Fourier series for the impulse train would appear to take one or other of the forms

$$\frac{1}{2\pi} \sum_{m=-\infty}^{+\infty} e^{imt} \quad \text{or} \quad \frac{1}{2\pi} \left\{ 1 + \sum_{m=-\infty}^{+\infty} \cos(mt) \right\}. \qquad (11.15)$$

Each of these series is divergent, and it is not immediately obvious in what sense either of them could be said to represent the periodic train of delta functions. A nonstandard treatment, such as that given below, can make the matter much easier to understand.

11.3.2 A pre-delta function of Dirichlet type. We recall first some elementary results from standard Fourier series theory:

Let $f(t)$ be any standard function which is continuously differentiable everywhere and periodic with period 2π. Then its Fourier series is well defined and converges to the value $f(t)$ for all $t \in \mathbb{R}$. That is,

$$f(t) = \frac{1}{2\pi} \sum_{n=-\infty}^{+\infty} c_n e^{int}$$

where

$$c_n(f) = \int_{-\pi}^{+\pi} f(t) e^{-int} dt.$$

This means that we can write

$$f(t) = \lim_{n \to \infty} \frac{1}{2\pi} \sum_{k=-n}^{n} c_k e^{ikt}$$

$$= \lim_{n \to \infty} \frac{1}{2\pi} \sum_{k=-n}^{n} e^{ikt} \int_{-\pi}^{+\pi} f(\tau) e^{-ik\tau} d\tau$$

$$= \lim_{n \to \infty} \int_{-\pi}^{+\pi} f(\tau) \left\{ \sum_{k=-n}^{n} e^{ik(t-\tau)} \right\} d\tau$$

$$\equiv \lim_{n \to \infty} \int_{\pi}^{\pi} f(t - \tau) D_n(\tau) d\tau,$$

where

$$D_n(t) = \frac{1}{2\pi} \sum_{k=-n}^{n} e^{-ikt} = \frac{1}{2\pi} \sum_{k=-n}^{n} e^{ikt} \equiv \frac{\sin((n + 1/2)t)}{2\pi \sin(t/2)}.$$

In particular we can take any smooth function f which vanishes outside $(-\pi, \pi)$ and get

$$f(0) = \lim_{n \to \infty} \int_{-\pi}^{\pi} f(-\tau) D_n(\tau) d\tau \equiv \lim_{n \to \infty} \int_{-\infty}^{+\infty} f(t) D_n(t) dt,$$

using the fact that $D_n(t)$ is even. Accordingly we conclude that the internal function $\Delta_{(2\pi, D)}(t)$ generated by the sequence $(D_n)_{n \in \mathbf{N}}$ is a pre-delta function on the interval $(-\pi, \pi)$ which, following a suggestion by Laugwitz, we may call a **pre-delta function of Dirichlet type**. We can get an explicit expression for this Dirichlet type pre-delta function by carrying out a perfectly legitimate infinite summation

$$\Delta_{(2\pi, D)}(t) = \frac{1}{2\pi} \sum_{k=-\Omega}^{\Omega} e^{ikt} = \frac{\sin((\Omega + 1/2)t)}{2\pi \sin(t/2)}. \tag{11.16}$$

This Dirichlet pre-delta function is periodic with period 2π for $t \in {}^*\mathbb{R}_{bd}$. That is, it behaves like an infinite periodic sum of local pre-delta functions concentrated at the points $2m\pi$, where m ranges over the hyperintegers from $-\Omega$ to Ω. If $f(t)$ is any continuously differentiable function which vanishes outside some finite real interval (a, b) then for some (finite) integers m_1, m_2 we surely have

$$\int_{-\infty}^{+\infty} \Delta_{(2\pi,D)}(t) {}^{*} f(t) dt \approx \sum_{m=m_1}^{m=m_2} f(2\pi m). \qquad (11.17)$$

But $\Delta_{(2\pi,D)}$ is not the same as the internal function $\Delta_{(2\pi,d)}$ defined in (11.13) and we cannot even write in general,

$$\Delta_{(2\pi,d)}(t) \equiv \sum_{m=-\Omega}^{+\Omega} \delta_{(d)}(t - 2\pi m) \approx \frac{1}{2\pi} \sum_{k=-\Omega}^{\Omega} e^{ikt} \equiv \Delta_{(2\pi,D)}(t).$$

$\Delta_{(2\pi,d)}$ and $\Delta_{(2\pi,D)}$ are equivalent only in the sense that they exhibit the same sampling properties at all points of the form $t = 2\pi m$, at least for all functions $f(t)$ continuously differentiable at such points. The relation between $\sum_{n=-\infty}^{+\infty} \delta(t - 2\pi n)$ and its "Fourier series" therefore has meaning only in some operational sense.

11.3.3 A pre-delta function of Fejer type. Suppose now that $f(t)$ is a function of period 2π which is merely continuous everywhere on $\mathbb{R}$, and not necessarily smooth. Then, although the Fourier series of $f(t)$ over $(-\pi, \pi)$ can be defined formally, there is no guarantee that it will converge. In other words the limit as $n \to \infty$ of

$$s_n(t) = \int_{-\pi}^{\pi} f(t - \tau) D_n(\tau) d\tau$$

may or may not exist. If it does not exist then a standard way to proceed is to consider a transformation of the infinite Fourier series into a **summable** form. This means that the transformation is to be such that if the original series converges then so also does the transformed series and to the same sum: if the original series diverges then the transformed series may nevertheless converge. In the so-called **Cesaro** method of summation the idea is to replace the partial sums $D_n(t)$ by their arithmetic means,

$$\sigma_n(t) = \frac{1}{n}\{s_0(t) + \ldots + s_{n-1}(t)\} \equiv \int_{-\pi}^{\pi} f(t - \tau) K_n(\tau) d\tau,$$

where we define the standard functions K_n by writing

$$K_n(t) = \frac{1}{n}\{D_0(t) + \ldots + D_{n-1}(t)\} \equiv \frac{1}{2n\pi}\left\{\frac{\sin(nt/2)}{\sin(t/2)}\right\}^2.$$

Then for f continuous (or even merely piecewise continuous) on $\mathbb{R}$, and of period 2π, it can be shown that

$$f(t) = \lim_{n\to\infty} \sigma_n(t) \equiv \lim_{n\to\infty} \int_{-\pi}^{+\pi} f(t-\tau)K_n(\tau)d\tau \qquad (11.18)$$

uniformly for all $t \in \mathbb{R}$. This means that we are dealing with yet another (periodic) pre-delta function which is defined by the sequence $(K_n)_{n\in\mathbb{N}}$:

$$\Delta_{(2\pi,K)}(t) = \frac{1}{2\pi\Omega}\left\{\frac{\sin(\Omega t/2)}{\sin(t/2)}\right\}^2, \qquad (11.19)$$

and which, again following Laugwitz, we shall refer to as the **Fejer** predelta function. $\Delta_{(2\pi,K)}(t)$ is a periodic pre-delta function of Dirac type, and can be given an equivalent "Fourier series" representation since

$$K_n(t) = \frac{1}{n}\sum_{k=0}^{n-1} D_k(t) = \frac{1}{n}\sum_{k=0}^{n-1}\left\{\sum_{m=-k}^{k} e^{imt}\right\}$$

and we can alter the order of summation to obtain

$$\Delta_{(2\pi,K)}(t) \approx \frac{1}{2\pi}\sum_{k=-\Omega}^{\Omega}\left\{1 - \frac{|k|}{\Omega}\right\}e^{ikt}. \qquad (11.20)$$

11.3.4 Summability methods other than that of Cesaro may be used. Accordingly, let $\Delta_{(2\pi)}$ represent an arbitrary, 2π-periodic, internal function of the form

$$\Delta_{(2\pi)}(t) = \frac{1}{2\pi}\sum_{k=-\Omega}^{\Omega} d_{|k|}e^{ikt} = \frac{d_0}{2\pi} + \frac{1}{\pi}\sum_{k=1}^{\Omega} d_k \cos(kt) \qquad (11.21)$$

and suppose that Δ does satisfy a sampling property over $(-\pi, \pi)$ which is applicable to all "sufficiently smooth" functions. Then since $\cos(t)$ is infinitely differentiable we surely have, for all finite values of k,

$$1 = \cos(k.0) \approx {}^*\!\!\int_{-\pi}^{\pi} \Delta_{(2\pi)}(t)\cos(kt)dt = d_k.$$

Hence a necessary condition for $\Delta_{(2\pi)}$ to have such a sampling property is that $d_{|k|} \approx 1$ for all $k \in \mathbb{Z}$.

If $f(t)$ is any standard function of period 2π which is such that the sampling property of $\Delta_{(2\pi)}$ applies then we have

$$f(t) \approx {}^* \int_{-\pi}^{\pi} \Delta_{(2\pi)}(\tau) f(t - \tau) d\tau = {}^* \int_{-\pi}^{\pi} \Delta_{(2\pi)}(t - \tau) f(\tau) d\tau$$

$$= \frac{1}{2\pi} \sum_{k=-\Omega}^{\Omega} d_{|k|} e^{ikt} \, {}^* \int_{-\pi}^{\pi} f(\tau) e^{-ik\tau} d\tau = \frac{1}{2\pi} \sum_{k=-\Omega}^{\Omega} c_k(f) d_{|k|} e^{ikt}. \quad (11.22)$$

where

$$c_k(f) = \int_{-\pi}^{\pi} f(t) e^{-ikt} dt.$$

The choice of a "summability method" for the Fourier series of a standard 2π-periodic function $f(t)$ is thus seen to correspond to the choice of a pre-delta function $\Delta_{(2\pi)}$ such that the sampling property is valid for $f(t)$ at the points $2\pi m$.

11.4 N.S.A. AND DISTRIBUTIONS

11.4.1 Nonstandard theories of Schwartz distributions. The account just given of pre-delta functions suggests that similar nonstandard treatments of other generalised functions should prove useful. In Chapter 8 a brief introduction to the theory of distributions is sketched, using the classical formulation due to Schwartz in which these generalised functions are defined as (continuous) linear functionals on a basic space, $\mathcal{D}(\mathbb{R})$ of test functions. There have been many attempts to offer a treatment of distributions which is based on more familiar (standard) material and which therefore might be simpler to understand. The most well-known of such alternative approaches are the sequential theories developed by Mikusinski, Temple and others and subsequently exploited to great effect in texts by Lighthill and Jones. The development of a general theory of distributions as equivalence classes of internal functions on $^*\mathbb{R}$ is an obvious corollary. Such a treatment was already envisaged by Abraham Robinson in his original text on NSA in 1966. Since then the nonstandard approach to distributions has been adopted, and extended, by a number of workers in the field, most notably by Detlef Laugwitz.

In terms of a simple ultrapower model of the hyperreals, such as that given in Chapter 10, the basic idea is to exploit the fact that any distribution μ can be expressed as a limit of a sequence of regular distributions (i.e. ordinary locally integrable functions) $\mu_n(t)$. This means that for any test function $\phi(t)$ in $\mathcal{D}(\mathbb{R})$ we would then have,

$$\mu(\phi) = \lim_{n \to \infty} \mu_n(\phi) \equiv \lim_{n \to \infty} \int_{-\infty}^{+\infty} \mu_n(t)\phi(t)dt. \qquad (11.23)$$

In particular, let $(\theta_n)_{n \in \mathbb{N}}$ be a delta sequence of infinitely differentiable functions, so that $\theta_n \in C^\infty(\mathbb{R})$ for all $n \in \mathbb{N}$. Then we may take

$$\mu_n(t) = (\mu \star \theta_n)(t) \equiv < \mu(\tau, \theta_n(t - \tau) >$$

so that the distribution μ can actually be represented in terms of a sequence of infinitely differentiable functions. Then the internal function $F_\mu = [(\mu_n)_{n \in \mathbb{N}}]$ will be a nonstandard representative of μ, the action of μ on an arbitrary test function ϕ being given by

$$\mu(\phi) = st \left\{ \left[\int_{-\infty}^{+\infty} \mu_n(t)\phi(t)dt \right] \right\}. \qquad (11.24)$$

For example, let $\theta : \mathcal{R} \to \mathbb{R}$ be any function in $\mathcal{D}(\mathbb{R})$ such that $\int_{-\infty}^{+\infty} \theta(t)dt = 1$. Then $\delta_{(\theta)}(t) \equiv \Omega \, {}^*\theta(\Omega t)$ is an internal function on ${}^*\mathbb{R}$ which vanishes outside monad(0) and which is such that for all test functions $\phi \in \mathcal{D}(\mathbb{R})$ we have

$$st \left\{ {}^* \int_{-\Omega}^{+\Omega} \delta_{(\theta)}(t) \, {}^*\phi(t)dt \right\} = \phi(0).$$

The *-derivative of $\delta_{(\theta)}(t)$ is well-defined on ${}^*\mathbb{R}$ as the internal function $\Omega^2 \, {}^*\theta'(\Omega t)$ and we have

$$st \left\{ {}^* \int_{-\Omega}^{+\Omega} {}^* D\delta_{(\theta)}(t) \, {}^*\phi(t)dt \right\} = -\phi'(0).$$

And so on for higher order derivatives of the delta distribution.

The set of all internal functions on ${}^*\mathbb{R}$ which are infinitely *-differentiable is an internal set of objects in the nonstandard universe which we denote by ${}^*C^\infty(\mathbb{R})$. In the simple ultrapower model described in Chapter 10 an internal function $F = [f_n]$ belongs to ${}^*C^\infty(\mathbb{R})$ if and only if $f_n \in C^\infty(\mathbb{R})$ for nearly all $n \in \mathbb{N}$.

It is easy to see that if F belongs to $^*C^\infty(\mathbb{R})$ then so also does its nonstandard derivative $^*DF = [Df_n]$, and it follows that if F is a nonstandard representative of a distribution μ then $^*DF$ is similarly a nonstandard representative of the distribution $D\mu$. We cannot, of course, identify a distribution μ with a single such internal function F_μ. In general there will be infinitely many nonstandard representatives of a particular such μ, each of which we shall refer to as a **pre-distribution**. This is all too clearly illustrated by the case of the delta distribution, δ, and the corresponding family of pre-delta functions $\delta_{(\theta)}(t)$. It therefore becomes necessary to introduce an equivalence relation for such internal functions:

Distributional equivalence. Two internal functions F and G will be called **distributionally equivalent** if an integer k and two internal functions Φ and Ψ exist such that

(i) $\Phi(t) \approx \Psi(t)$ for all finite t, and

(ii) $F = {}^*D^k\Phi$ and $G = {}^*D^k\Psi$.

Then each standard Schwartz distribution μ will admit a nonstandard representation as an equivalence class $[F]$ of internal functions in $^*C^\infty(\mathbb{R})$.

However, for a satisfactory nonstandard theory of distributions there still remains a technical problem. It is not difficult to see that there also exist internal functions in $^*C^\infty(\mathbb{R})$ which are not pre-distributions. For example, if $\delta_{(\theta)}$ is any pre-delta function, then the pointwise product

$$\delta_{(\theta)}^2(t) \equiv \delta_{(\theta)}(t).\delta_{(\theta)}(t)$$

is certainly a well-defined internal function in $^*C^\infty(\mathbb{R})$ but it is equally certainly not a nonstandard representative of any Schwartz distribution. To complete the theory then, we would need to characterise those internal functions among the members of $^*C^\infty(\mathbb{R})$ which do behave as pre-distributions. A particularly straightforward and simple way of doing this is to make use of the work done by the Portuguese mathematician **Sebastiao e Silva** who, as remarked at the end of Chapter 8, developed a categorical axiomatic theory of Schwartz distributions. This Silva approach to distributions is both simple and elegant, and it makes only modest demands on the background in analysis which needs to be assumed of the student meeting the subject for the first time. It is still comparatively little known, but a very clear introductory account (in

English translation) by **J. Campos Ferreira** has recently become available. We give a very brief summary of the basic ideas in the next section, primarily to show how it can be used to derive a satisfactory nonstandard theory of distributions.

11.4.2 The Silva axioms. The crucial feature of distributions which justifies their description as generalised functions is surely their role as derivatives of ordinary functions in a generalised sense. Indeed, it can be shown that every Schwartz distribution is locally a finite order derivative of a continuous function. More particularly, every Schwartz distribution of finite order is globally a finite order derivative of a continuous function. For simplicity in what follows we shall confine attention to the case of such distributions of finite order, and use the Silva axioms to establish a simple nonstandard representation of them.

First, as primitive ideas on which to base the axioms themselves we have only to take the familiar concepts of **continuous function** and **derivative**. Then the Silva axioms can be stated as follows:

Axioms for finite-order distributions.

SA1. If f is continuous on $\mathbb{R}$, that is if f is a member of the linear space $C^0(\mathbb{R})$, then f is a distribution. We will often denote this distribution by the symbol μ_f.

SA2. To each distribution μ there corresponds another distribution, called the **derivative** of μ and written as $D\mu$, such that if $\mu = f \equiv \mu_f$, where f has a classical continuous derivative f', then $D\mu = f' \equiv \mu_{f'}$.

SA3. To each distribution μ there corresponds a non-negative integer k such that $\mu = D^k f \equiv D^k \mu_f$, for some $f \in C^0(\mathbb{R})$.

SA4. For any functions (distributions) f, g in $C^0(\mathbb{R})$ we have $D^k f = D^k g$ if and only if $(f - g)$ is a polynomial of degree less than k.

The most familiar model for this set of axioms is the classical Schwartz theory, as described in Chapter 8, in which a distribution on $\mathbb{R}$ is identified with a (linear and continuous) functional on the space $\mathcal{D}(\mathbb{R})$ of all infinitely differentiable functions of compact support. Then in accordance with axiom SA1 each function $f \in C^0(\mathbb{R})$ is identified with the (regular) distribution μ_f which is defined as the functional given by the equation

$$< \mu_f, \phi > \equiv \int_{-\infty}^{+\infty} \phi(t) f(t) dt \qquad (11.25)$$

for all $\phi \in \mathcal{D}(\mathbb{R})$.

The derivative of a distribution μ in the Schwartz model is defined by setting

$$< D\mu, \phi > \equiv < \mu, -\phi' > \qquad (11.26)$$

again for all functions $\phi \in \mathcal{D}(\mathbb{R})$.

Examples.
(i) The unit step function u appears in the Schwartz model as the functional defined directly on $\mathcal{D}$ by

$$< u, \phi > = \int_{\infty}^{+\infty} \phi(t) u(t) dt \equiv \int_0^{+\infty} \phi(t) dt.$$

In the axiomatic formulation it is simply defined as the (distributional) derivative of the continuous function t_+, which is such that $t_+(t) = t$ for all $t \geq 0$ and which vanishes for all negative values of t.
(ii) The delta distribution δ is usually defined as that functional which maps each function $\phi \in \mathcal{D}(\mathbb{R})$ into the number $\phi(0)$:

$$< \delta, \phi > = \phi(0) >$$

It can be identified in the above axiomatic formulation as the (distributional) second derivative of a continuous function by means of the equation

$$\delta(t) = D^2 t_+(t). \qquad (11.27)$$

(iii) Similarly the pseudofunction $t_+^{-3/2}$ can be defined as a distribution in the Silva formulation as the second order (distributional) derivative of the continuous function $-4u(t)\sqrt{t}$.

11.4.3 A nonstandard model for the Silva axioms.

In order to describe a nonstandard model for the Silva axioms we need to review the concept of continuity as it applies to internal functions. In section 10.4.6 we have already defined the property which results from a transfer of the criterion for standard continuity to the nonstandard universe:

*-continuity An internal function $F : {}^*\mathbb{R} \to {}^*\mathbb{R}$ is *-continuous at $a \in {}^*\mathbb{R}_{bd}$ if and only if given any positive hyperreal ξ (which may possibly be infinitesimal) there exists a corresponding positive hyperreal $\eta = \eta(\xi)$ such that

$$\forall t \in {}^*\mathbb{R}_{bd} : |t - a| < \eta \Rightarrow |F(t) - F(a)| < \xi.$$

Internal functions which are *-continuous will inherit nonstandard analogues of characteristic properties of standard functions which are continuous in the usual, standard sense.

S-continuity. If, on the other hand, we simply apply the standard definition of continuity directly to internal functions we obtain another form of continuity:

An internal function $F : {}^*\mathbb{R} \to {}^*\mathbb{R}$ is said to be **S-continuous** at $a \in {}^*\mathbb{R}_{bd}$ if and only if given any positive real number r there exists a corresponding positive real number $s = s(r)$ such that

$$\forall t \in {}^*\mathbb{R}_{bd} : |t - a| < s \Rightarrow |F(t) - F(a)| < r.$$

A necessary and suficient condition for F to be S-continuous at a is that $F(t) \approx F(a)$ for all t such that $t \approx a$.

A standard function $f : \mathbb{R} \to \mathbb{R}$ is continuous in the usual sense at a real point $a \in \mathbb{R}$ if and only if its nonstandard extension ${}^*f$ is S-continuous at a. Also, if f is any internal function which is S-continuous at a real point a then its standard part ${}^oF(t) \equiv st\{F(t)\}$ is a function which is continuous in the standard sense at A. Every internal function F in ${}^*C^\infty(\mathbb{R})$ is *-continuous on $\mathbb{R}$ but not necessarily S-continuous. We shall denote by ${}^SC^\infty(\mathbb{R})$ the set of all $F \in {}^*C^\infty(\mathbb{R})$ which are finite-valued and S-continuous at each point t of $\mathbb{R}$.

If f is any standard function which belongs to the space $C^\infty(\mathbb{R})$ of all infinitely differentiable functions then it is certainly true that its nonstandard extension ${}^*f$ is a member of ${}^SC^\infty(\mathbf{kR})$. This will not be so if f is merely a continuous standard function. However what we can say is that given any such continuous function f on $\mathbb{R}$ we can always find an internal function F in ${}^SC^\infty(\mathbb{R})$ such that $F(t) \approx {}^*f(t)$ for all t in $\mathbb{R}$. Hence every regular distribution generated by a continuous function has nonstandard representatives (pre-distributions) in ${}^SC^\infty(\mathbb{R})$.

Next, we denote by ${}^SD^\infty(\mathbb{R})$ the space of all internal functions F in ${}^*C^\infty(\mathbb{R})$ such that F is a finite-order *-derivative of some function Φ in

$^{S}C^{\infty}(\mathbb{R})$. Then this space $^{S}D^{\infty}(\mathbb{R})$ contains all those functions which are nonstandard representatives of finite-order distributions on $\mathbb{R}$; that is, it contains all (finite-order) pre-distributions. Every finite-order Schwartz distribution can then be identified with an equivalence class of such pre-distributions, formed with respect to the criterion of distributional equivalence given above.

Further reading.

Abraham Robinson is undoubtedly the key figure in the creation of what has come to be called Nonstandard Analysis, and he pioneered much of its subsequent application to many fields of mathematical analysis. In particular he developed the first nonstandard theory of Schwartz distributions, including of course the delta distribution. But as regards the delta function itself this was substantially anticipated in a remarkable paper by **C.Schmieden** and **D.Laugwitz** published in 1958. Moreover the subsequent detailed exploitation of nonstandard treatments of the delta function has been very considerably due to Laugwitz, and the treatment described in this final chapter is based essentially on his work. Much of what Laugwitz acheived is only available in publications in research journals and deserves to be much more widely known since it represents some of the clearest and most approachable treatments of the subject area to have been developed so far. Some idea of the breadth of Laugwitz's contribution can be found in **Advances in Analysis, Probability and Mathematical Physics: Contributions to Nonstandard Analysis**, (Kluwer, Academic Publishers, 1995), edited by **S.A.Albeverio, W.A.J.Luxemburg** and M.P.H.Wolff.

The very brief description of a nonstandard treatment of Schwartz distributions based on the Silva axiomatic formulation which follows later in the chapter is intended to indicate what might be done to place the whole Schwartz theory on a conceptually simple footing, in which all distributions admit (like delta functions) representations as functions in the proper sense of the term. The analysis given here is presented in much greater detail in the book by **J.Sousa Pinto** referred to in the notes on **Further Reading** at the end of Chapter 10. This book also contains extended application of nonstandard methods to harmonic analysis and generalised Fourier transforms.

Supplementary accounts together with nonstandard theories of ultradistributions and the more complex problems presented by Sato hyperfunctions and the so-called New Generalised Functions of Colombeau will be found in **Theories of Generalised Functions**, by **R.F.Hoskins** and **J.Sousa Pinto**, Horwood Publishing, 2005. This book is a revised and updated version of the original text entitled **Distributions, Ultradistributions and Other Generalised Functions**, Ellis Horwood Series[, 1994.

Solutions to Exercises

CHAPTER 2

Exercises I.

1. (a) $u\{(t-a)(t-b)\} = 0$ for $t \in (a,b)$, 1 for $t \notin [a,b]$.

(b),(c) $u(t-e^{\pi}) = u(t - \log \pi) = 1$ for $t > \log \pi$, and $= 0$ for $t < \log \pi$.

(d)
$$u(\sin t) = \begin{cases} 1 & \text{for } 2\pi n < t < (2n+1)\pi \\ 0 & \text{for } (2n-1)\pi < t < 2n\pi \end{cases}$$

(e)
$$u(\cos t) = \begin{cases} 1 & \text{for } (2n-1/2)\pi < t < (2n+1/2)\pi \\ 0 & \text{for } (2n-3/2)\pi < t < (2n-1/2)\pi \end{cases}$$

(f) $u(\sinh t) = u(t)$; (g) $u(\cosh t) = 1$ for all t.

2. (a) $\text{sgn}(t^2 - 1) = +1$ for $|t| > 1$, and $= -1$ for $|t| < 1$.

(b) $\text{sgn}(e^{-t}) = 1$, for all t.

(c)
$$\text{sgn}(\tan t) = \begin{cases} +1 & \text{for } n\pi < t < (2n+1)\pi/2 \\ -1 & \text{for } (2n-1)\pi/2 < t < n\pi \end{cases}$$

where $n = 0, \pm 1, \pm 2, \ldots$

(d)
$$\text{sgn}(\sin 1/t) = \begin{cases} +1 & \text{for } 1/(2n+1)\pi < t < 1/2n\pi \\ -1 & \text{for } 1/2n\pi < t < 1/(2n-1)\pi \end{cases}$$

where $n = 1, 2, \ldots$. Also, $\text{sgn}(\sin 1/t) = +1$, for $t > 1/\pi$ and to complete the definition for negative values of t we need only note that the function is odd.

(e) $t^2 \operatorname{sgn}(t) = t^2$, $t > 0$, and $= -t^2$, $t < 0$;

(f) $\{\operatorname{sgn}(\sin t)\} \sin(t) = |\sin t|$.

(g) $\sin t\{\operatorname{sgn}(\cos t)\} = \sin t$, $(4n-1)\pi/2 < t < (4n+1)\pi/2$, and $= -\sin t$, $(4n+1)\pi/2 < t < (4n+3)\pi/2$.

3. (a) $u_{1/2}(t)$; (b) $u_{1/e}(t)$.

4. (a) $-3t^2\operatorname{sgn} t$; (b) $e^{|t|}\operatorname{sgn} t$; (c) $-e^{-|t|}\operatorname{sgn} t$; (d) $\operatorname{sgn} t.\cos t$; (e) $\operatorname{sgn}(\sin t).\cos t$; (f) $\{\operatorname{sgn}(\sin(t))\}.\cos t$; (g) $\operatorname{sgn} t.\cosh t$.

5. (a)$f(t_0 + h) - f(t_0) = h\{f'(0) + \epsilon\}$, where $\epsilon \to 0$ with h.

(b)
$$f'(t) = 2t\sin(1/t) - \cos(1/t), \quad t \neq 0; \quad f'(0) = 0.$$

(c)
$$g'(t) = 2t\sin(1/t^2) - \frac{2}{t}\cos(1/t^2), \quad t \neq 0; \quad g'(0) = 0.$$

6. Let $g(t) = tu(t)$; then $g(b) - g(a) = \int_a^b u(t)dt$.

Exercises II.

1.
$$\int_{-\infty}^{+\infty} f(t)\delta(-t)dt = \int_{-\infty}^{+\infty} f(-\tau)\delta(\tau)d\tau = f(0).$$

2.
$$\int_{-\infty}^{+\infty} \left\{ \frac{u(t-a+\eta) - u(t-a)}{\eta} \right\} dt = \frac{1}{\eta}\int_{a-\eta}^{a} f(t)dt,$$

which $= f(\xi)$ for some ξ such that $a - \eta < \xi < a$.

3. Since $u(-t) = 1 - u(t)$ we have $\{u(t)\}' = -\delta(t)$. Similarly since $\operatorname{sgn}(t) = u(t) - u(-t)$, we have $\{\operatorname{sgn}(t)\}' = 2\delta(t)$.

Exercises III.

1. $\int_{-\infty}^{+\infty} \cos(t)d_n(t)dt = 2n\sin(1/2n) \to 1$.

2. $\int_{-\infty}^{+\infty} \cos(t)g_n(t)dt = 2n^2\{1 - \cos(1/n)\} \to 1$.

3. $\int_{-\infty}^{+\infty} \cos(t)h_n(t)dt = 2\{4n^2\cos(1/2n) - 2n^2[1 + \cos(1/n)]\}$, which tends to 1 as $n \to \infty$.

4. (a) $\sum_{r=1}^{n} f(t_r)\Delta_r\phi = f(t_1)\{u_c(t_1 - a) - u_c(0)\} = f(t_1)(1 - c)$, for every partition of $[a, b]$ so that $\int_a^b f(\tau)du_c(\tau - a) = f(a)(1 - c)$. The argument for (b) is similar.

CHAPTER 3

Exercises I.

1. (a) $\cos(t)$; (b) $\sin(t) + \delta(t)$; (c)$1 + 2e\delta(t - 1)$.

2. (a) 5; (b) 1/3; (c) e^{-4}; (d)$\sinh(4)$; (e)$\pi + 1/2e^2$;

   (f) $\sum_{k=1}^{n} k^2 = \frac{1}{6}[n(n + 1)(2n + 1)]$; (g) 0; (h) $2e^{\pi}$.

Exercises II.

1. (a) $\delta(t + 1) - \delta(t - 1)$; (b) $\{u(t) - 1\}\sin(t) - \delta(t)$;

(c) $\{u(t - \pi/2) - u(t - 3\pi/2)\}\cos(t) + \delta(t - \pi/2) + \delta(t - 3\pi/2)$;

(d) $e^4\delta(t + 2) + \delta(t) - e^{-4}\delta(t - 2)$

$$-2e^{-2t}\{u(t + 2) + u(t) - u(t - 2)\}.$$

2. (a) $\text{sgn}(t)$, $2\delta(t)$; (b) $-e^{-|t|}\text{sgn}(t)$, $e^{-|t|} - 2\delta(t)$;

(c) $\text{sgn}(t).\cos(|t|)$, $2\delta(t)-\text{sgn}(t).\sin(t)$.

3. $\lim_{t\to0+} f(t) = +1$ and $\lim_{t\to0-} f(t) = -1$

$f'(t) = -1/[t^2\cosh^2(1/t)]$, $Df(t) = 2\delta(t) - 1/[t^2\cosh^2(1/t)]$

Finally note that

$$\lim_{t\to0} 1/[t^2\cosh^2(1/t)] = \lim_{x\to\infty} x^2/\cosh^2(x) = 0.$$

Exercises III.

1. (a) $-b, 2ab, b^3 - 3a^2b$; (b) 1; (c) 19.

2. $\int_{-\infty}^{+\infty} f(t)\delta'(-t)dt = \int_{-\infty}^{+\infty} f(-\tau)\delta'(\tau)d\tau = +f'(0)$.

3. $\frac{d}{dt}\{\phi(t)\delta'(t)\} = \frac{d}{dt}\{\phi(0)\delta'(t) - \phi'(0)\delta(t)$

and $\phi'(t)\delta'(t) + \phi(t)\delta''(t)$

$$= \{\phi'(0)\delta'(t) - \phi''(0)\delta(t)\} + \{\phi(0)\delta''(t) - 2\phi'(0)\delta'(t) + \phi''(0)\delta(t)\}.$$

Exercises IV.

1. (a) $-\frac{1}{5}\sinh(4/5)$; (b) $\frac{1}{\pi}\cosh(\pi^2)$; (c) $2\cosh(\pi/2)$;

(d) $\frac{1}{\pi}\left\{1 + 2\sum_{m=1}^{\infty} e^{-m}\right\} = \frac{e+1}{\pi(e-1)}$.

2. (a) $\sum_{n=1}^{\infty}\{\delta(t - n\pi) + \delta(t + n\pi)\}$; (b) $\frac{2}{\pi}\sum_{m=-\infty}^{+\infty}\delta(t - (2m+1))$.

(c) $u(e^t) \equiv 1$. Hence $\frac{d}{dt}u(e^t) = 0$ and so $\delta(e^t) = 0$.

(d) $\frac{1}{2\pi}\left\{\frac{\delta'(\theta-\pi)}{2\pi} - \frac{\delta(\theta-\pi)}{2\pi^2} - \frac{\delta'(\theta+\pi)}{2\pi} + \frac{\delta(\theta+\pi)}{2\pi^2}\right\}$; (e) $\frac{1}{4}\delta(x)$.

3. $\delta(x - 1)$; $\delta(x - a)$.

CHAPTER 4

Exercises I.

1. (a),(b), (c) linear; (a), (d), (e) time-invariant.

2. $T[x] + T[-x] = T[x - x] = 0$. Hence $T[-x] = -T[x]$. Let $x = m/n$, where m and n are positive integers. Then

$$T[mx] = mT[x] \quad \text{and} \quad nT[x/n] = T[n(x/n)] = T[x].$$

3. $x(t - a) \rightarrow \int_{-\infty}^{+\infty} x(\tau)h(t, \tau + a)d\tau$ which, by time-invariance, is equal to $y(t - a) \equiv \int_{-\infty}^{+\infty} x(\tau)h(t - a, \tau)d\tau$. Hence $h(t, \tau + a) = h(t - a, \tau)$, so that $h(t, a) = h(t - a, 0)$.

Exercises II.

1. $\sigma(t) = tu(t)$; $h(t) = t\delta(t) + 1u(t) = u(t)$.

2. (a) $(u \star u)(t) = tu(t)$; (b) $\frac{\alpha e^{\alpha t}}{\alpha^2 + \omega^2}$; (c) $\{\frac{t}{2}\sin(t)\}u(t)$.

(d) $(p \star p)(t) = 1 - |t|$ for $|t| \leq 1$, and $= 0$ for $|t| > 1$.

3. $g_n'(t) = g_{n-1}(t)$ and $g_n(0) = 0$ so that $g_n(t) = \int_0^t g_{n-1}(\tau)d\tau$ for $n \geq 1$.

4. $\int_0^t \frac{d\tau}{\sqrt{\tau}\sqrt{t-\tau}} = \int_0^{\sqrt{t}} \frac{2x\,dx}{x\sqrt{t-x}} = \pi$, so that $(g \star g)(t) = u(t)$.

Exercises III.

1. 0 (since $\sqrt{2} + \sqrt{3}$ is a root of $x^4 - 10x^2 + 1 = 0$).

2. (a) A; (b) e^{-sa}; (c) s; (d) $1/s$.

3. $h(t) = u(t)a\cos(at)$; $y(t) = \frac{at}{2}\sin(at)$, for $t > 0$.

CHAPTER 5

Exercises I.

1. $\frac{1}{s}$; $\frac{1-e^{-s}}{s}$; $\frac{1}{s-a}$; $\frac{1}{s^2}$; $\frac{s}{s^2-b^2}$; $\frac{\omega}{s^2+\omega^2}$; $\frac{s}{s^2+\omega^2}$.

2. $y(t) = \{t + \cos(t) - 3\sin(t)\}u(t)$.

3. $y(t) = \{1 + 2t\}u(t)$.

4. $x(t) = \{5e^{-t} + 3e^{4t}\}u(t)$, $y(t) = \{5e^{-t} - 2e^{4t}\}u(t)\dot{\iota}$

5. $(g \star g)(t) = u(t)$ so that $G(s)G(s) = 1/s$.

Exercises II.

1. (a) $\delta''(t) - \delta'(t) + \delta(t) + e^{-t}u(t)$;   (b) $\frac{1}{2}\{\delta(t) + \delta(t-2)\}$;

(c) $\sum\limits_{k=0}^{n-1} \delta^{(k)}(t-k)$.

2. (a) $\frac{1}{s}\left\{\frac{1-e^{-as}}{1-e^{-\pi s}}\right\}$;   (b) $\frac{1}{2s}\left\{\frac{1-2e^{-as}e^{-s\pi}}{1-e^{-\pi s}}\right\}$.

3. $e^{-\pi s/2}/\{1 - e^{-\pi s}\}$.

Exercises III.

1. (a) $\frac{s^2-2a^2}{s(s^2-4a^2)}$;   (b) $\frac{1}{2}\left\{\frac{1}{s+1} - \frac{s+1}{s^2+2s+5}\right\}$;

(c) $\frac{3}{(s-1)^4} + \frac{3}{(s+1)^4}$;   (d) $\frac{s^2-a^2}{(s^2+a^2)^2}$;   (e) $\frac{6as^2-2a^3}{(s^2+a^2)^3}$.

2. Put $g(t) = f(t)/t$ so that $f(t) = tg(t)$ and $F(s) = -G'(s)$.

(a) $\arctan(1/s)$;   (b) $\frac{1}{2}\log\left\{\frac{s^2+b^2}{s^2+a^2}\right\}$;   (c) $\frac{1}{s}\arctan(1/s)$.

3. (a) $\frac{1-e^{-s}-se^{-s}}{s^2}$;   (b) $\frac{1-(s+1)e^{-s}}{s^2(1-e^{-s})}$;   (c) $\frac{1-(s+1)e^{-s}}{s^2(1-e^{-2s})}$;

(d) $\frac{1}{s^2} - \frac{e^{-s}}{s^2} - \frac{e^{-2s}}{s}$;   (e) $\frac{1-e^{-s}-se^{-2s}}{s^2(1-e^{-2s})}$;

CHAPTER 6

Exercises I.

1. If $f(t)$ is even then $f(t)\sin(\omega_0 t)$ is odd; hence $B_n = 0$. Similarly if $f(t)$ is odd then $f(t)\cos(\omega_0 t)$ is odd, and $A_n = 0$.

2. (a) $2\sum_{n=1}^{\infty}(-1)^{n+1}\frac{\sin(nt)}{n}$; (b) $\frac{\pi}{2} - \frac{4}{\pi}\sum_{n=1}^{\infty}\frac{\cos(2n-1)t}{(2n-1)^2}$;

(c) $\frac{\pi^2}{3} + 4\sum_{n=1}^{\infty}(-1)^n\frac{\cos(nt)}{n^2}$.

Finally, set $t = \pi/2$ in each of the above series.

3. $\cos(xt) = \frac{2x\sin(x\pi)}{\pi}\left\{\frac{1}{2x^2} + \sum_{n=1}^{\infty}(-1)^n\frac{\cos(nt)}{x^2-n^2}\right\}$,

for $-\pi < t < +\pi$. This is continuous at $\pm\pi$; put $t = \pi$ and divide by $\sin(x\pi)$.

4. (a) $\cos(t) = \frac{8}{\pi}\sum_{n=1}^{\infty}\frac{n}{(2n)^2-1}\sin(2nt)$;

(b) $\sin(t) = \frac{2}{\pi} - \frac{4}{\pi}\sum_{n=1}^{\infty}\frac{\cos(2nt)}{(2n)^2-1}$.

Exercises II.

1. $\frac{kd}{T}\left\{1 + \sum_{n=1}^{\infty}\frac{\sin(n\omega_0 d/2)}{n\omega_0 d/2}\cos(n\omega_0 t)\right\}$, $nT < t < (n+1)T$.

For $d = T$ all Fourier coefficients vanish apart from the constant term; $f_T(t) = k$ for all t. For $d = T/2$ we have

$$f_T(t) = \frac{k}{2}\left\{1 + \frac{4}{\pi}\sum_{n=1}^{\infty}(-1)^{n+1}\frac{\cos(2n-1)\omega_0 t)}{2n-1}\right\}.$$

Finally, if $k = 1/d$ and we allow d to tend to zero then $f(t) \to \delta(t)$ and $\frac{\sin(n\omega_0 d/2)}{n\omega_0 d/2} \to 1$.

2. (a) $\frac{1}{2\pi}\left\{1 + 2\sum_{n=1}^{\infty}\cos(nt)\right\}$;

(b) $\frac{1}{2\pi}\left\{1 + 2\sum_{n=1}^{\infty}(-1)^n\cos(nt)\right\}$;

(c) $\frac{2}{\pi} \sum\limits_{n=0}^{\infty} \cos(2n+1)t$.

4. $\sum\limits_{n=-\infty}^{+\infty} \delta\{t - (2n+1)\pi/2\} = \sum\limits_{n=-\infty}^{+\infty} \delta\{(t-\pi/2) - n\pi\}$

$= \frac{1}{\pi} \sum\limits_{n=-\infty}^{+\infty} (-1)^n e^{i2nt}$

$D^2|\cos(t)| = 2 \sum\limits_{n=-\infty}^{+\infty} \delta(t - (2n+1)\pi/2) - |\cos(t)|$, so that

$$C_n = \frac{2(-1)^n}{\pi(1-4n^2)} = \frac{2(-1)^{n+1}}{\pi(2n-1)(2n+1)}.$$

Exercises III.

2.

$$h_3(t) = \begin{cases} t + (a+b)/2 & \text{for } -(a+b)/2 \le t < (a-b)/2 \\ b & \text{for } (a-b)/2 \le t < (b-a)/2 \\ (a+b)/2 - t & \text{for } (b-a)/2 \le t < (a+b)/2 \end{cases}$$

3. (a) $2/(1+\omega^2)$;   (b) $-2i\omega/(1+\omega^2)$;   (c) $\sqrt{2\pi}e^{-\omega^2/2}$;

(d) $4(\sin(\omega) - \omega\cos(\omega))/\omega^3$:   (e) $\frac{1+e^{-i\omega\pi}}{1-\omega^2}$;   (f) $\frac{2\cos(\omega\pi/2)}{1-\omega^2}$.

5. (a) $e^{-|t|} = \frac{2}{\pi} \int_0^{+\infty} \cos(\omega t)d\omega \int_0^{+\infty} e^{-\tau}\cos(\omega\tau)d\tau$

$= \frac{2}{\pi} \int_0^{+\infty} \frac{\cos(\omega t)}{\omega^2+1}d\omega$. Hence $\int_0^{+\infty} \frac{\cos(xt)}{x^2+1}dx = \frac{\pi}{2}e^{-t}$.

(b) $1 - t^2 = \frac{2}{\pi} \int_0^{+\infty} \cos(\omega t)d\omega \int_0^1 (1-\tau^2)\cos(\omega\tau)d\tau$

$= \frac{4}{\pi} \int_0^{+\infty} \cos(\omega t) \left\{ \frac{\sin(\omega) - \omega\cos(\omega)}{\omega^3} \right\} d\omega, \quad |t| < 1$.

Put $t = 1/2$ to get $\int_0^\pi \frac{\omega\cos(\omega) - \sin(\omega)}{\omega^3} \cos(\omega/2)d\omega = -3\pi/16$.

Exercises IV.

1. (a) $\pi\delta(\omega) + \frac{1}{i\omega} - \frac{1}{a+i\omega}$;

(b) $\pi\delta(\omega) + \frac{\pi}{2}\{\delta(\omega - 2a) + \delta(\omega + 2a)\}$:

(c) $\pi\delta(\omega) - \frac{\pi}{2}\{\delta(\omega - 2a) + \delta(\omega + 2a)\}$.

2. $\hat{F}\left[\displaystyle\sum_{k=0}^{2N-1}\delta(t-kT)\right] = \displaystyle\sum_{k=0}^{2N-1}e^{-ik\omega T} = \dfrac{1-\exp(-i\omega T2N)}{1-\exp(-i\omega T)}$

$= \dfrac{\sin(N\omega T)}{\sin(\omega T/2)}\exp\{-i\omega T(N-1)/2\}.$

3. $f(t)\displaystyle\sum_{n=-\infty}^{+\infty}\delta(t-nT) = \dfrac{1}{T}\displaystyle\sum_{n=-\infty}^{+\infty}f(t)e^{in\omega_0 t}$, where $\omega_0 \equiv 2\pi/T$.

The Fourier transform of $f(t)e^{in\omega_0 t}$ is $F(i\omega - in\omega_0)$, and the Fourier transform of $F(nT)\delta(t-nT)$ is $f(nT)e^{-in\omega T}$.

4. $\hat{F}\{f_{(*)}(t)\} = \dfrac{1}{T}\left\{F(i\omega) + \displaystyle\sum_{n-1}^{+\infty}F(i\omega \pm i2\pi n/T)\right\}$

If $F(i\omega)$ vanishes outside $(-\pi/T, +\pi/T)$ then the translates $F(i\omega \pm i2\pi n/T)$ do not overlap. Hence

$$H(i\omega)\hat{F}\{f^*(t)\} = \frac{1}{T}F(i\omega),$$

so that the output is $\dfrac{1}{T}f(t)$. Since

$$h(t) = \frac{1}{2\pi}\int_{-\infty}^{+\infty}H(i\omega)e^{i\omega t}d\omega = \frac{1}{T}\left\{\frac{\sin(\pi t/T)}{\pi t/T}\right\}$$

it follows that $f_{(*)}(t) \star h(t) = f(t)$.

Index

Printed in the United States
By Bookmasters